DR. GARRETT L. TURKE

THERE'S A WINDOW TO HEAVEN

17 DAYS OF REVELATION AND HOPE

D0910575

ISBN 978-0-9970195-1-3

Printed by CreateSpace, An Amazon.com Company. Available from Amazon.com, CreateSpace.com, and other retail outlets.

Lead Editor:
Lindsay Marie Brumgard

Cover and book interior design:
Marie Kar – www.redframecreative.com

Cover photos:
Photo of sunburst over Lake Leelenau, Michigan, taken by Dr. Garrett Turke

Sillhouette photo by Maksim Evdokimov © 123rf.com

DEDICATION

There's a Window to Heaven recounts 17 consecutive days of near death, out of body events my father experienced while in the throws of advanced Alzheimer's disease and diagnosed with a rapidly escalating, terminal lung infection.

I dedicate this, my second book, to my children, Miah and Shani, who had the extraordinary experience to witness my father's apparent journey to the next realm of life. Additionally, a heartfelt dedication belongs to my nieces, Lindsey and Brityn, also present during those experiences, and so influenced have gone on to nursing careers.

To all of my "adopted" children, Adam, Antriece, Amadou, Aicha, Amelia and Aurelio: I have been blessed to know you during my time on this earth! I also dedicate to you this work, for you have provided me with continued hope and inspiration. For the very first time I am noticing that all of your names begin with the letter "A." I am struck, rather profoundly, by that "coincidence."

Continued love and admiration goes out to my mother, Rosemarie, and my sister, Tiffanie, without whom I would never have chosen to create a new career path for myself. Mom, your story is coming next.

Also, a special thanks to my brother-in-law Mike. For without your support for my mother and sister, I would not have been in position to write this book.

The deepest thank you is given to Lindsay, who took this book and breathed life into it before it went to typesetting. I am indebted to your wonderful editing and support for <u>There's a Window to Heaven</u>, and hope this experience will catapult you into your own professional writing career.

Finally, a loving dedication is also extended to my close friend, Donald Jenks, Jr., the father of my nieces, Brityn and Lindsey. In 2010, Don was pronounced dead from cardiac arrest following severe atrial fibrillation during heart valve replacement surgery. Resuscitation was attempted for 45 minutes as he lay lifeless, "flat lining," and without oxygen. He was, by standards of modern science, dead. The doctors' pronouncement to our family that Don had died was interrupted by an excited critical care nurse, who suddenly burst into the somber consultation room.

"We got him back!"

As I write this, Don is preparing to have coffee with me and begin his own book-writing journey. He too, had similar ethereal "visits" and experiences as my father.

I love and cherish all of you more than I could ever express.

— Garrett

"Dad," "Uncle Garrett," "Mr. Owl," and "Gertrude" (LOL Don!)

May 2018

SPECIAL ACKNOWLEDGMENTS

A special thank you and acknowledgment goes out to Antriece, who helped me guide <u>There's a Window to Heaven</u> to all the blessings I have since received. I am so grateful that you have reconnected with me after all these years. I can still remember, vividly, the day I took your training wheels off your bike when you were six! God bless you and your family.

I would also like to thank my longtime friend and colleague, Richard, for his unwavering support and belief in me. That phone call changed my life. Thank you for teaching me the true meaning of Grace.

And finally, a special acknowledgment to Amadou. I have always believed in you, and was so glad to help you with your education. You have, in return, given me the very best present, calling me "Dad." Love and blessings to you and your family.

WITH RESPECT

There's a Window to Heaven was not written to further an agenda, cause, or the ideals of any particular faith. As a person who has traveled widely, I have been exposed, at times quite intimately, to many of the world's religions. I have been profoundly impacted by these faiths throughout my life. I believe in God, in whatever form that God appears.

Over the years I have, for one reason or another, "adopted" many children, most of whom who have grown to be adults by the time this book was written. They have all either stayed in touch with me or found their way back to me. My "children" come from many faiths and have different beliefs of life and death. In my mind, I cannot favor one over the other by personally ascribing to a specific faith.

"By belonging to none I belong to all…"

Garrett

PROLOGUE

In late July 2007 my father Walter, already deep in the throws of a seven-year battle with Alzheimer's disease, was hospitalized with an impacted bowel. Subsequent CAT scans also uncovered a serious lung infection and pneumonia. The infection insidiously worsened over the course of a week, with antibiotics seemingly powerless to slow its progression. Further x-rays showed both lungs to be collapsed and filled with hundreds of infection-filled cysts. As a family we were offered the option of surgery, which would at worst kill him and at best result in a long painful recovery with more cognitive loss. We decided to take him off all life sustaining medical supports, manage his pain with medication, and let him pass peacefully. Doctors gave him no more than two to three days to live.

He didn't die. His language, which had been all but lost for years, came back.

For the next 17 days my father described "trips" he was taking to what he called a "shining city filled with light," where there was "no fighting, only love." Before each "visit," he described the coming of "angels" which he traced in an arc over his head, followed by the arrival of "a kind man with a grey beard." This experience was followed by what looked like an electrical "jolt" and my father's reports that he was "flying," with the ability to look down upon us in his room.

These experiences occurred daily, sometimes multiple times, for over two weeks, while in the hospital and when returned to his memory care "home" on hospice. The experience was witnessed by scores of people, including nearly all members of my family and our children, hospital staff, and care-giving staff at his care facility. As one of his nurses told us, "We see these events in many dying patients; only your dad has a lot more of it. We used to tell the doctors but they never believed us, so we have just stopped talking about it with them."

There's a Window to Heaven recounts the 17 days of what many others have called "near-death" or "out of body" experiences. The explanations as to the cause of these seemingly universal experiences vary from hallucinations induced by the dying process to actual visits to an afterlife. In the case of my father, there were events he described which make me question scientific explanations of neurologically induced hallucinations or other medical phenomena. These events made me question the outer parameters of scientific knowledge and what we think we know of the human experience.

As a psychologist and "man of science," I fretted for years about whether to tell the story of my father's out of body experiences outside of a small circle of trusted others. I wrote an entire book, 497 Nails, about my thirteen-year journey taking care of my father, without ever including these two-plus weeks of ethereal experiences. Maybe I was not ready. Maybe I was afraid of being discredited, ridiculed, or ostracized. Certainly, I am aware these social consequences may still occur.

This morning I woke up no longer afraid, or at least able to override my self-generated fears. It is time to tell this part of my father's story.

"17 Days"

17 days to say goodbye
17 days to really mean it
17 days until the end
17 days until a new beginning

—*Miah Joy Turke*
August, 2007
my daughter, age 11 at the time

"I am sending an angel ahead of you,
to keep you on your journey,
and to guide you to a place
I have prepared."

Exodus 23:20

INTRODUCTION

It was the last week of July, 2007. I was enjoying two weeks of vacation with my wife Stacy, and our two daughters, Shani, then age 18, and Miah, age 11. The weather had been beautiful ever since we arrived, the waters of northern Lake Michigan warm, and the days still long. We had been able to get away from the insidiously worsening stress of taking care of my beloved father, Walter, stricken with Alzheimer's in early 2000. He was now requiring round the clock care in a "memory care facility" and facing a long, unrelenting, downward spiral toward the end of his time on this earth.

My sister Tiffanie had graciously offered to give us some respite for a couple weeks. Frankfort, Michigan in the summer was as close to heaven on earth as one could get, the sun streaming off turquoise blue shimmering waters that kissed up to pristine white and ever shifting sand dunes. While the world thought we were on vacation, I was there on another mission: I needed to reclaim my wife, my kids, and my family. They had been neglected. My dad had eaten up so much time. I had to make up for the debt owed my family, a debt emanating from the unbalance created by an ever-so-slowly dying parent.

Midway into our vacation, Tiffanie called. It was just a voice mail that sat there annoyingly but seemingly harmlessly for a while. I just didn't want any distraction from my vacation. "Hey Garrett, its Tiff. Hey everything's OK down here, but just so you know Dad was taken to the hospital today. Its OK, please don't worry, we got it covered. They just think he has an impacted bowel and is having trouble passing stool. He probably will just be in overnight. Just thought you should know. Now go back to having fun, I've got this covered."

In the realm of all reasons a 77 year-old man with Alzheimer's could be hospitalized, I guess this should be considered good news. It was easily fixable and it wasn't my watch. Tiff didn't sound mad or even distressed. "I got this," she said. Yeah this is no problem, I can get back to our vacation with minimal accountability, minimal stress, and minimal guilt. That's just what I did. I filled Stacy in with business-like precision, didn't tell the kids, and continued with plans to go on a canoe trip that would take us from the Platte River into the waves of open Lake Michigan.

Tiff left another voice mail the next day. This one wasn't perceived as annoying, and I told Stacy immediately. "Hey Stacy, got a call from Tiff, Guess it's the all clear; Dad probably just got released from the hospital. I'm just gonna let it be."

"Honey, I think you should listen to it. It's probably good news...but you can never tell when someone's in the hospital." She was speaking from experience. Her mother had passed away a decade before after a seemingly benign hospitalization for back spasms. She caught a staph infection while in the hospital from an IV port. The infection found its way to her heart muscle, and eventually killed her. That unexpected, shocking loss had shut Stacy down for nearly five years.

I put the phone to my ear. "Hey Garrett, its Tiff again. We are at the hospital. Hey they got his bowel problem cleared up, but they also found a lung infection on the CAT scan. Don't worry, if it's pneumonia they will be able to treat it, but he will probably be here a couple more days. Not to worry, I got this. Lindsey and Brityn are helping and everything's fine. Dad seems pretty happy actually, with all this attention. So anyway, go back to your family, I'll just send you an update here and there."

Two days passed without a word more. I assumed things were fine, and I was having a great time with Stacy and the kids. For the first time in what seemed like years, we had truly managed to get away. I had found my family again. It was good to have them back as my priority.

At 11:30-something on a sunny Thursday morning, Tiff left a third voice mail. We were getting ready to go to the beach, so I let the voice mail sit there, convinced and even relieved by my perception of its innocence. Hours later, when we returned, we found a note taped to the front door of our rental. It was from the resort's office. "Dr. Turke, your sister is desperately trying to reach you. Please call immediately!"

My heart could surely be heard pounding over the sound of waves crashing back at the shoreline. What the hell? Pneumonia is treated pretty routinely in the hospital. Maybe Tiff is just overreacting. I punched in the numbers, lucky to get the sequence right because of my adrenalin rush, a familiar feeling since Dad had gotten sick. The robotic yet serene generic female voice said I had three voice mails, all from the same number: Tiffanie's number.

"Garrett I am at the hospital. 6th floor, Room 604. I am so sorry. Please come back! We need you here, I need you." Tiff was crying. "Dad's lung infection is worsening, he's got a 103 fever with medicine, they can't seem to stop the infection. Garrett, please, I am sorry. His lungs are collapsing oh my God!"

Two hours later we had packed up the All-American Chrysler mini van. The vacation had to end abruptly, another family event once again trumped by Dad's escalating needs. I thought Stacy and the kids would be upset. They had been having the time of their lives, yet neither Stacy, or the girls showed any real sign of distress, emotional upset, disappointment, or anger. *Tough resilient kids and family*, I thought, but that was not it. They were simply used to this. Dad's needs always found a way to override whatever life we were living.

Five hours later the turquoise blue water, big body-surfing waves, and the pristine white dunes were a distant memory, becoming smaller and smaller in the rear view mirror, then gone. We arrived home, and dumped our bags, toys and surfboards in the garage. The kids and Stacy went inside to have dinner. I pulled my clothes bag out of the pile and threw it back in the van. "See you guys, I'm going to catch up with Tiff at the hospital."

Over the next three days I was with Dad, Tiffanie, and every combination of our kids, spouses, and extended family. Dad's condition had worsened each day. The doctors and nurses couldn't keep his fever down. He wasn't responding to antibiotics. Getting restless, he kept signaling with his dementia-saddled remnants of language that he just wanted to go home. Wheezing, and with labored coughs, he could no longer breath comfortably. There was a horrible smell that couldn't be described. My mind gave it an explanation I dared not speak I could only say it inside my mind, *It is the smell of death.*

On Dad's fifth day in the hospital, a Monday, the doctors broke the morning with news that both Dad's lungs were collapsed or nearly collapsed, and filled with hundreds of multiplying grape-sized pustules. The infection was relentlessly worsening, impervious to a whole cocktail of antibiotics. With his immune system wearing out, and his strength weakening, the doctors seemed to be at a loss. There was talk of surgically opening him up and scraping out the infection, a life threatening invasion that could potentially kill any elderly person, and even more so, one stricken with advanced Alzheimer's.

Dad was dying, by all medical criteria. A host of specialists came into his room that morning, including a pulmonary specialist, an infectious disease specialist, a radiology specialist, and finally a social worker. They all seemed to already be at an inevitable conclusion. We called our family together, discussed the pros and cons of invasive surgery versus just letting our tired father and patriarch pass. We decided to sleep on it, delaying what we all secretly knew was going to be our answer.

Dad even had a voice in the matter. He had spoken to us many years ago, long before he got sick. His words were now just a memory, but they rang so loud in my head they couldn't be ignored. "Garrett, if I ever get so sick that I am going to be in pain, and have no real quality of life, please don't let me be a burden to you and your sisters. Just let me go, and you guys go on with your lives. I have already donated my body to the medical school. It's all paid for, they will take care of my body, the transport, the details, everything, its all been taken care of. Please don't have a funeral, that's so morbid, maybe some day you can have a memorial service for me and rejoice that I was once in your life."

Tiffanie wanted a miracle and some sort of herculean recovery, which by all the doctors' reports was not going to happen. He was already dying. We had to accept that this was the end. Dad's selfless decision, made when he was still healthy, could not be ignored. I told Tiffanie that maybe this was God's way of sparing Dad another 5 years of further erosion from Alzheimer's disease. There seemed to be not much left for him to lose anyway. He had virtually no language left, didn't seem to recognize us as family anymore, and needed assistance in every facet of life including eating. Now his immune system was failing. The future course was unavoidable, the collision with the end inescapable. It would be traumatically painful to let Dad go. Would it also be a relief, a release from an incredible burden, a burden Dad once anticipated and didn't want us to carry?

ROOM 604

There was no good news on Tuesday morning, Dad's sixth full day in the hospital. Even with fever-reducing medication, his fever persisted between 101 and 103 degrees. Although he couldn't communicate it, my father seemed to be in pain much of the time. He would grimace while awake and while asleep, and once shouted at two nurses trying to reposition him to prevent bedsores. His cough made a defeated kind of sound, unproductive and useless to clear any airways. Again, that smell, that horrible smell of infection oozing from every exit point in Dad's body, was inescapable. Although I tried not to think about it, I kept thinking what it would be like to drown, or suffocate to death.

Tiffanie and I had postponed our decision about Dad's future until today. We had discussed our options, or lack thereof, with everyone yesterday and decided to sleep on it. Neither Tiffanie nor I, the two powers of attorney assigned years ago by Dad, had slept at all. Today was to be the day we let go, after years of propping him up, praying for miracles, and wading through a veritable flood of denial. The floodwaters were now finally draining away, receding, leaving behind just a frail remnant of a man I used to call The Champion. Dad was no longer even in there, in any recognizable form.

We had to make the formal decision, a decision we had already made in our heads, that we weren't ready to speak. To subject him to a future of chronic pain, in order to appease our own selfish fantasies, was unacceptable. This was it. We all knew this day would eventually come, and had readied ourselves for it. Today we learned, however, that any amount of self-believed readiness could not even come close to adequately preparing us. To be legal, the decision had to be verbalized, and we were the ones that had to verbalize it. Saying it out loud was the hardest thing I had ever had to do.

Dad's family physician, Dr. Jeffries, would be in soon, and he would likely be the executor of our decision. We would go with hospice, and probably just let him pass away in the hospital. The doctors said at his rate of escalating infection, he was not going to make it more than a day or two. To try to move Dad anywhere would be extraordinarily confusing, disorienting and painful for him, so we just decided to let him go where he was at, in room 604.

Dr. Jeffries came in. He looked at us compassionately, rare for doctors these days, opening his eyes wide as if asking a question without saying a word.

"We've decided, Doctor Jeffries, to spare Dad of any unnecessary pain." I choked on the words. Tiffanie cried, hard. Dad was asleep, seemingly oblivious to our declaration of his death sentence.

Dr. Jeffries knew that not much more needed to be said. "I will go tell the hospital staff."

Before Dr. Jeffries could leave the room I verbalized the second most difficult thing I have ever had to push out. "Dr. Jeffries, how is he going to pass, I mean, he has an escalating lung infection, I can see he is having trouble breathing, is he going to suffocate? I can't imagine anything more painful than that."

Dr. Jeffries tried to be as kind as he could. "He will pass peacefully, we will see to that. We will manage his pain, with morphine most likely, and he will pass from either a lack of oxygen or sepsis, which means the infection has gotten into his bloodstream. But don't worry; when his body no longer gets enough oxygen he will become unconscious, and go into a coma. We will see that he doesn't have any pain."

Dr. Jeffries smiled warmly and patted my shoulder. It was probably the most he could do without violating any doctor-patient-family boundaries. Maybe that was all he could do anyway. I thought, *My God, he probably has had to have this speech hundreds of times with hundreds of families. How could anyone do this over and over again?*

Dr. Jeffries soon left the room. After days of endless waiting for test results, Tiffanie and I expected we had some time before Dr. Jeffries' orders would be delivered, let alone enacted. Not so. Within ten minutes a nurse was in Dad's room. She unhooked his IV, discontinued all antibiotics, and took away most of his electronic monitors. This was it. Tiffanie let out a gut-wrenching sob and hugged me so hard I couldn't breathe for a moment. We were told that hospice would be in to see us later today, that we would meet a coordinator, either Sue or Cara, and a chaplain named Steve from the Heartfelt Hospice group we had selected.

Then, in the excruciating quiet of the hours waiting for the arrival of hospice and the return of our family members, "it" began. Startling "jolts" of what I could only call "electricity" shivered up and then back down Dad's body in his hospital bed. Then again it happened… and again.

We had seen these jolting occurrences before, earlier in the year, and thought it was some type of seizure activity. We had Dad evaluated and he was once taken him to the ER overnight for medical tests. All results were inconclusive. One explanation was that he was showing an idiosyncratic reaction to his medication regimen. Another hypothesis was that areas of his brain were so compromised by his Alzheimer's disease that his neurological circuitry was "misfiring." Dad's Alzheimer medications were changed as a result, and the "jolts" stopped. Now these strange episodes were back. We had no explanation why, other than his neurological system was beginning to "misfire" again.

Dad's jolts occurred episodically for the rest of the afternoon. They went on for a number of minutes each time, seemingly traveling the length of his body. Then, the jolting would stop, sometimes for many hours before resuming again. This episodic but increasingly familiar activity continued periodically through the night. Dad was seemingly oblivious, and calm in between episodes. He was sometimes asleep and sometimes awake when they occurred. When we reported these episodes to the doctors and nurses, they offered that these "jolts" were probably related to his system starting to shut down, a precursor to his passing. Tiff and I took it as a sign that Dad wouldn't be with us much longer. We sat there, numb, barely talking, watching the jolts as if they were some kind of fait de complete.

Hospice eventually came in; we met with Sue and signed a bunch of paperwork. They told the Michigan State University School of Medicine that Dad's body would be arriving this week. Chaplain Steve came in and read the story of the Good Samaritan, which I had indicated was Dad's favorite parable. By all accounts, the end was near. Our family and extended family arrived throughout the late afternoon and evening, including our children. Very little was said other than just some hugs. We all knew it was the last goodbye. There was some relief in this I think, as Dad's Alzheimer's had already created what seemed to be an endless series of goodbyes over the past seven years.

Tiff and I stayed with Dad all night. We wanted to be there when the end came. Dad's jolting continued through the night and persisted into the next morning. I encouraged Tiffanie to go home and tend to her family for a few hours. Dad apparently would be with us one more day, but we were resigned that there would be no miracle. Not this time.

The next day, a Wednesday, Dad's jolts would be accompanied for the first time by another occurrence, an event I would come to call "the visits."

THE VISITS: DAY 1

I sat by Dad's bedside pensively, making up my obsessive-compulsive list for the day. It was a day that I was certain would be Dad's last. Even though Dad was going to pass soon, there was work to be done, and people to inform. It was almost if Dad was already dead. Conscious of how precise and dutiful I was in making my list, I tried to punish myself for being callous and insensitive. I reassured myself. Dad's disease had prepared me well for this moment, years ago. I had done everything a good son should do.

I was alone with Dad. Having shooed Tiffanie out in the wee hours of the morning, I was hoping she could try to get some much-needed sleep and tend to her family. We had been told that Dad's death, though looming now, would still be a "process," and that it wouldn't happen abruptly. When things worsened, we would have plenty of time to assemble Dad's troops at his bedside. All of us, to a person and including our young children, wanted to be at his bedside when the end came.

Dad lay rather unremarkably a mere eighteen inches away from me, propped up at what looked like an uncomfortable 45-degree angle on his hospital bed. Just a basic vitals monitor was still connected to him. One IV port was left in place, in order to deliver his pain medications, which were all derivatives of opium. I had a stray thought about how little difference there was right now between a person on pain medication and a heroin user. They would each have the same high. Dad was going to go out completely numb. He would be completely oblivious to the world around him as he made his exit.

At 7:40 AM, the jolting started to happen again. I just noted the time and wrote it down on my pad. I was no longer startled by these occurrences. Once fearing they were seizures, I used to immediately call for medical help and would feel my heart race. Now, they were such a common event that I no longer reacted. *Dad's electrical circuitry is misfiring*, I thought. The next thought escaped my mouth, though barely audible. "What a goddamn, wicked disease."

I watched an electrical shiver run the length of Dad's body, from feet to head and back down, then encompassing his whole body at once. This was followed by the expected, a pronounced jolt. Dad tried to sit up, opened his eyes wide, and spoke an intelligible sentence. That startled me, for he had not produced an intelligible sentence in well over a year. His disease had mercilessly taken his language a long time ago.

"The man, ah, maybe he's my guide, well he's my friend for sure. He has a name, but either he can't remember or it's unimportant."

What the hell????? I sat up as straight as an arrow and pulled my chair to the edge of Dad's bedside. "Dad?"

"Yeah I am going to ask him if it is all right to tell you about him, and tell the others."

"Who is he, Dad?"

"I don't know. He has a name but he won't tell me. He says it's not important."

"What does he look like?"

"He is a kind man, you can see it in his eyes. He has a gray beard. He can fly!"

"Does he have wings?"

"No, no, no. He is not that kind of a person! There are others with wings but he is by himself. For God's sake, Garrett, he is right here, can't you see him?" Dad hadn't called me Garrett in two years. The word, "son," had disappeared long before that.

"He is going to take me somewhere…just a minute, I will ask him if you can come with us?"

For a second, I had lost all sense of reality, space and time, startled and afraid by Dad's offer. "No, no, that's OK Dad, I am fine right here." Apparently in an effort to salvage some bizarre measure of self-respect, I still offered to remain involved in Dad's apparent hallucination and delusion.

"Ask him if I can take notes about what you tell me?" My self-induced railing began immediately. *That's right Garrett, go right into your comfort zone, your safe place of being a psychologist, presenting the illusion of intimacy while remaining safely distant.*

"He says that will be fine. But you must tell the others about this."

"OK." I felt strange about participating in this nonsensical excursion, but was intrigued...or maybe enticed. I wondered for a second who "the others" were. I assumed it was our family. I would later find out this was only partially correct.

"He wants you to do it, to write it down."

"OK."

"But you must tell the others."

"Sure. I will write it down and then tell the others. So, how are you feeling, Dad?"

"I am well when I am with him, you know, not like I am now. He is showing me how to fly."

"What?"

Dad actually chuckled, like he thought my lack of understanding was amusing. "How else do you think we will get there?"

An orderly named Kathy came in. She watched our dialogue for a moment, as she was checking the status of the room. I told her that Dad was apparently hallucinating, and shared a little about what he was saying.

Kathy was completely unfazed. "He is not hallucinating."

Dad started to drift off, away from us. He was mumbling almost unintelligibly, but I could make out some of the words. They were all numbers. "6,5,5,4,4,4... " Maybe Kathy's appearance had broken up the connection.

I returned to Kathy's statement that Dad was not hallucinating. "Of course he's hallucinating, he's talking about some man taking him somewhere. He is talking about being well, not sick, and flying!"

Kathy was still unfazed. "He is not hallucinating. Many, many people have this experience. I see it all the time here. My husband died last year. He talked about this too."

I could still hear Dad faintly. "4,4,4, 3, 3..." It was like he was counting down. He was touching his left fingers with his right index finger as he counted.

"What do you think is happening?" I peered straight into Kathy's eyes. They looked kind, tired and worn by years of life, but honest and sincere.

"Well if you want my thought..."

"Yes please, I do..."

"He is showing you a window to the next world, a window to heaven."

THE VISITS: DAY 2

Dad had had a quiet night, at least in terms of being able to sleep, however fitfully. Tiffanie and I were pretty much working in several hour shifts, with each of us not wanting to be far away, either in time or distance. Dad looked somewhat comfortable, with his array of pain meds, though he wheezed and coughed and strained when he breathed. He seemed to be getting enough air in. I was surprised his fever hadn't spiked and stayed up high, as we had been told to expect that. Still running a temperature of 101 to 103, his infection was not backing down, however. The episodic jolts, what the doctors called "shivers," were continuing, and were occurring independently of his state of consciousness. We thought each 12-hour block of time would be his last. All our family members who lived nearby were at the ready to come down at a moment's notice.

Dad's ability to speak had seemingly returned to the early stages of his Alzheimer's diagnosis. Certainly not the psychiatrist-level cognition he once commanded, but he was getting out intelligible sentences. Many of the nurses said this was common for people who were about to pass, even for those with Alzheimer's. There could be bursts of lucidity that could last from hours to a couple days. The nurses stressed, however, that family members often misconceived these moments as signs of a miracle recovery. We were told to not get our hopes up.

About 10 AM, Dad started experiencing another "electrical" jolt and discharge, with the familiar shaking or shivering running from feet to head and back down. He was doing this daily now, both awake and asleep, and we were used to it. This time, however, there also was a strong startle reaction. Dad's eyes opened wide. I helped him as he tried to sit up.

"They're here!"

"Who's here, Dad, the man with the grey beard?"

"No, no, no, not him, he always comes last. I mean them, the ones that can fly. You see them right here."

Dad counted off numbers like he did the previous day, but this time he was pointing to an arc shaped line in the air over his head. "Lets see, one, one, two, two, there's three, so one two three, there's four, yes four, and now five. There's five today!"

"You've seen them before?"

"Yes, yes, they come every time, before he comes, the one who has a name but who says it's not important."

"What do they do?"

"I'm not really sure. They don't talk, but I can hear them. They are just for me, its like they are getting me ready for my trip. Yes, yes, that's their job. They are always happy. And they are just for _me_!"

"Oh….wait, he's here, he's here! The man is now here."

"The man with the grey beard?"

"Yes, he's come to take me to the special place. They are all waiting for me. I'll be gone two days this time I think!" Dad sounded happily excited, like a little kid at Christmas.

"I'm gonna go up, way up, I will be so high but I can look down and see you."

"He said that after today I will be able to get there, by myself, with the others."

"Where we are going, I've been there many times now, it's a very shiny, very bright city. Very, very bright! And we are always happy. I can see my parents, and oh yes my sister, yes Friedi is there too. People are singing and dancing. There is no conflict, nobody is ever hurt, no fighting." Dad made an umpire-like baseball gesture for being safe, as if to emphasize his point even more.

I had heard and read that some people with near-death or out of body experiences often described a realm where they elevated above their surroundings, and the presence of an all enveloping, soothing white light or brightness. I also knew modern scientists usually attributed these experiences to hallucinations that were spawned by neurochemical changes in the brain, assumed to emanate from changes occurring within a dying brain. Yet these images that Dad was describing were, so far, nearly the same every time, and quite vivid. His descriptions were always so consistent. As a psychologist, my previous experiences with others having hallucinations caused by psychosis were never consistent like this.

"Well the man says I will be gone for two days, I guess I will see you when I get back. But I am already trying to tell him its hard to come back, I mean, well I am sorry but I don't really want to leave after I've been there."

With that Dad closed his eyes and appeared to go to sleep. About an hour later, two of my closest work colleagues, JoAnn and Ruth, arrived. JoAnn had attended one of my photography shows a couple years before and purchased an image. I had taken Dad with me to deliver the art a few days later, driving to JoAnn's country home, about an hour away. They sat down quietly so as not to disturb Dad's sleep.

"How's it going, Garrett? We are so sorry for your situation and your dad. A lot of people at work are praying for you and wanted to come, but we didn't want to overwhelm you or your dad." JoAnn's piercing, nearly black eyes told me how heartfelt these words were. There was then a long, but not uncomfortable, period of silence. When someone is dying there's not much to say.

Ruth then asked, making small talk to break the silence. "So how is your dad doing?"

"Well, his fever's pretty constant at 101 or 102, up to 103 sometimes, his lungs are filled with hundreds of infected pustules, and both lungs are collapsed or nearly collapsed. He is expected to die from pneumonia or sepsis within a day or two. We took him off all the life support; he just has pain medicine now. They said it won't be painful, he'll go into a coma and then just slip away without distress." I felt myself choke up on the last few words and couldn't get all the syllables out. JoAnn put her arms around my shoulders and pulled me in a little.

I collected myself, seemingly trying to be stoic in front of my work colleagues. They knew me more as Dr. Turke, the Court's "venerated psychologist," an image I was never all that comfortable with. Now I was just Garrett, the more real and fragile son of a dying father.

"Well, something else has been going on, it's very strange. Dad's language has come back, and he's been telling me about these trips he's been taking, to this place filled with light."

Ruth pulled her chair in, and then leaned on the seat's edge, almost into me.

"Yes, I don't know why I am telling you this, I am sure its just one of those hallucinations you sometimes hear about when people are dying. "

"Describe what your dad is telling you, Garrett. I mean, please, can you?"

Well he says he goes to this city that glows and is all bright and shiny. He says he sees his father and mother there, and that they live there, along with his sister Frieda. All these people are deceased; his sister was younger than him but died about 40 years ago in a car accident."

"He says he flies there, and there are these beings that have wings and that help take him, or navigate for him, or something like that. And there's another being that comes especially for him. He has a grey beard and is very kind. The man seems to be like a guide."

"When they get to the place its described like a city, with people singing and dancing, and no fighting or conflict of any kind."

Ruth wasted no time. An avid and observant church-going Christian, but certainly not a zealot, she said confidently, "This is described in the Bible, exactly as your dad describes. He is in heaven, Garrett!"

"Yes I know. Look, I am writing all this down." I held up my increasingly filled legal pad, as if to validate what Ruth was saying. But I wasn't all that confident in my belief. The scientist in me saw to that. "I must believe something is going on...but it could be just hallucinations, too."

Ruth shook her head, humbly rejecting my ambivalence or skepticism. "How many of these visits has he had? And is it the same each time?"

"Yeah, it's exactly the same each time, which doesn't fit with the typical hallucinations I have seen mentally ill people have. And he seems totally serene, totally calm.

You'd think if they were true hallucinations they wouldn't be so clear and vivid, or as consistent, and they would be scary to him."

"Before he goes wherever he goes, it looks like an electrical current is entering him. The medical staff calls them shivers. I am telling you, they are not shivers. I already know its like electrical energy entering and then leaving, but no one seems to hear me."

"The holy spirit is entering him." Ruth was now staring at my dad, seemingly to glean any trace of further evidence.

JoAnn just stood there, quiet.

Then came another episode. I thought, *Wow; these aren't just for my family to witness*. A strong surge of what I was determined to call electricity permeated my father's body. He shook for a few seconds, and then opened his eyes. He looked straight at JoAnn and smiled. "I...I...I know you!"

Dad continued, smiling. Dad didn't even recognize his family anymore much less someone he met just once, a long time ago when I was delivering some artwork.

Dad continued, "I... I... I've met you, twice!"

Ok, now I knew Dad was unintentionally fabricating this, he's seen JoAnn only once before. I corrected him, rather insensitively, by talking to JoAnn. "He probably only thinks he met you, he does this with people who are kind with him, it's his Alzheimer's. He's only met you that one time when we delivered your image. I really don't think he remembers."

JoAnn interrupted me by talking with my father. "No, your son Garrett is the one who's got it wrong. You met me twice, once at Garrett's photography show, you were there, you smiled at me, and then again when you two dropped off my print."

Dad grew excited. "And we drove and drove down that bumpy dirty road!"

"Yes, that's right Walter, I live on a long, dirt road and it gets soooo muddy. I can never keep my car clean."

Ruth couldn't hold back. "So Walter, Garrett says you go on these trips to this city?" She wanted even more evidence that provided confirmation for her belief system.

"Yes. I was just there for two days." I looked at my watch, which was sort of ridiculous. Of course he hadn't been gone two days. He had been "gone" exactly 43 minutes.

"Yes, I go to the shiny city. I know how to get there now. They can go with me, but I don't need them anymore really. It's easy to find. Just up, up, up and then wow, we are there already."

"How old are you there?" I asked, maybe beginning to believe some more, well at least just a little.

"I am young, maybe 35, like my sister who is there. She lives there I think, with my mother and father. They are still alive!"

"Are there people with wings, you know, like are there angels?" JoAnn was now joining Ruth's quest for confirmation.

"Well there are these people, they sort of are like us, but yes they have wings and they can fly all the time. I don't have any wings when I go, but I am flying too. Yes, you finally said it; I didn't know what to call them. They are angels. My angels!"

"There is singing and dancing, it's all lit up, shining, like the sun is out all the time, but a lot more…oh and I saw Connie there this time. You know my friend Connie, from home?" Dad now looked over at me, for his own validation I think. Indeed, there were two Connies at his memory care facility, one staff and one resident.

"You mean Constance, the one from East Africa?"

"No. no, no. Not her!" Dad sounded incredulous, like why don't you get it? "That Connie is way too young. And she's my girlfriend. I am talking about Connie, you know the one, who is like my grandmother?" Although the same age, Dad had an Alzheimer's induced habit of seeing same aged peers as always being much older.

"Yes Dad, now I know, Connie, the one who lives there."

"Now you're finally making sense! Yeah I saw her there, too. I wonder what she was doing there? Maybe just visiting, like me."

Ruth continued with her interview. To her, there was no longer any trace of ambivalence that heaven was real. "So, Walter, this man with the beard, who is he and what does he do?"

"I've asked for his name, as has the Doctor over here," pointing now at me. "He won't tell me. He says he has a name but it's not that important. He's there every time. With the angels, though he usually gets here last."

"And this place he takes you to, it's always good, I mean friendly and kind, you feel like you belong?" Ruth peered into Dad's wide eyes.

Dad now looked incredulously at Ruth. He pointed at JoAnn, gesturing for her to come closer. He then whispered, "Its so much better than that!" Dad chuckled, like Ruth had missed the mark.

JoAnn had brought Dad some ice cream, hoping he would be allowed to eat a little. Dad thanked her but then looked straight at her, "I don't need food anymore."

Then Dad made a sweeping motion with his hands; as if to say, I am including all of you, pay attention here! He started crying a little, then stammered out some words... "I, I didn't know...(long pause)... I didn't know if.... I didn't know if I, if I...."

I intuitively finished his sentence. Years of trying to communicate through the Alzheimer's haze had taught me well.

"If I...."

"Believed in God?"

Dad eyes grew huge. "That's it! Now you've got it and you understand! I didn't know if I believed in God and now... I know!" Dad seemed immensely proud, with a child-like giddiness.

The afternoon crept on by. JoAnn and Ruth left, and Dad went back to sleep. Tiffanie would be out soon. She had seen Dad have these experiences already, but I couldn't wait to tell her some more. How on earth did Dad suddenly remember JoAnn? For years now, he had only been able to sustain about a 5-minute memory span.

At 4:40, before Tiffanie and her family arrived, Dad woke up, although just for a minute or two.

His eyes opened, he looked straight up at the ceiling, and said, "And I am working on something for tomorrow."

THE VISITS: DAY 3

Tiffanie had made me go back home to get some sleep. I had resisted, fearing I would miss Dad's last moments, but she used the same words I had used to shoo her out for some rest. "You need to detach, get some rest, and take care of your family." But I had lost my grounding. Suddenly, words like "normal," "reality," "time," and "rational" had lost their meaning. I felt usurped from the familiar, like it had been taken from me and I had been transported to some "other world."

After a fitful night, I finally got up and looked at the sunrise out over our backyard pool. I noticed a strange, black, ominous-looking "growth" on the bottom of the pool's deep end, near the drain. It would later turn out to be a form of algae that could be eliminated with pool chemicals, but I didn't see it that way. I took it as a "sign." I stared at the spot, which was about five feet in area, wide at the top and becoming increasingly narrow toward the base. It had the shape of a tornado, alive, almost swirling. Though irrational, somehow it scared me immensely.

When I got to the hospital mid afternoon on Day 3 of "the visits," there had already been a lot of activity in Dad's room. Heartfelt Hospice had been in touch with Dad's care facility, and had decided that Dad had met met all the criteria for being on end of life hospice supports. It was agreed that Dad would be transferred back to Claire Bridge, his memory care facility and focal point of everything familiar to him. Dad had lost touch with who everyone was, where he was, what his life had been, but he remembered Claire Bridge. I reasoned this was because my father felt universally and unconditionally loved there, something that had eluded him in his own childhood, and maybe his whole life.

Although plans had been made to send Dad back to Claire Bridge as soon as possible, there seemed to be all sorts of bureaucratic signatures and release forms and doctors' sign-offs that needed to be completed before anything could happen. I had seen all this before, and told my sister that his transition probably wouldn't happen until tomorrow or even Friday. I thought to myself, *Why even bother; he'll be dead by tomorrow anyway.* The loving, sensitive part of my brain had already given way to the practical, lets-just-get-this-over-with side. I felt a twinge of guilt, but it wasn't very powerful. Dad's death had been rehearsed in my mind a thousand times over during the past several years. I secretly hoped this was indeed the end, as I had had enough. I felt as if I had earned the right to be selfish, after everything I had been through taking care of him.

Tiff was the first to show visible signs of distress. "Garrett, I can't believe he's made it this long, he was supposed to have passed two days ago. I don't think he'll make it until tomorrow if this hospice thing takes that long. He hasn't had any food in three days, and hardly any water now for the past two. Oh my God! Why won't God just take him? He is breathing so hard, oh my God I don't want him to suffer any more pain!"

I patted Tiffanies shoulder, a fake, cold, 21st century attempt at intimacy. "Be strong, Tiffanie. That's all we can do."

Dad aroused a little in his bed, after being asleep and "ethereally inactive" for the morning. His eyes suddenly opened wide, as if startled. "They're here!" Dad had a big smile on his face. He traced an arc over his head, and began counting, as we were now accustomed to. "One, two, three, lets see that's three, oh yeah now four, and five! There's five today! They are all here, for me!"

"Do you know who they are now, Dad?" I asked. Dad's face looked aglow. He no longer seemed sick, but had the gaze of an excited traveler. "They are my…my…my friends. And the good man will be here soon, I just know it." A rush of jolts careened up and down Dad's spine, again confirming to me that whatever was happening had to be electrical. What else could it be? I wasn't ready to say he was being "possessed" or taken over, like by some other entity. The scientist in me just said quietly, *it's some kind of electrical activity.*

Dad exclaimed excitedly, "I am going to the city of light, the shining place, where everyone is so happy. There is no fighting, only love! They will be singing! And I will be there soon! My parents are there, waiting for me!"

Just then my curiosity, and bourgeoning faith, if you want to call it that, was interrupted by a voice. I heard it clearly, but it seemed to be coming from the inside of my head. I tried to make it out. It was definitely an adult male voice, but it didn't sound like mine. "Garrett, you need to listen." It scared me. This was no longer my little science experiment with my dad. This was starting to unnerve me. I couldn't get the picture of that "tornado" that was forming on the bottom of the pool out of my head. I was soon enveloped by a massive anxiety attack.

I looked at Tiff and scared as a skittish rabbit, I bolted out of the room. Without saying anything I walked as fast as I could down the hall and made a left turn into a brightly lit vestibule, then into a long corridor of the hospital. It was a sunny day, and the sun's rays were streaming through the many windows and skylights.

"Garrett...you need to listen." *Jesus God, Jesus God! Leave me alone! Where are you coming from?*

I could clearly hear this freakin' voice but it was on the inside. I looked around and saw hospital staff and family of patients walking by. They couldn't hear anything. *Oh my God, oh my God, I am loosing my mind! I can't handle this, I can't handle this!*

The voice came back again. I moved further down the corridor, but "it" followed me.

Finally I decided to pray, but it really was more like desperate pleading. *Please God, I don't know what you want with me or my dad, this is scaring the shit out of me, please could you just let me deal with this without panicking anymore, I am going to crack up! I have to be strong for everyone, my dad, my kids. I am so sorry I don't go to church more, I am so sorry I don't belong to anything, now I want to belong and its too late. It's too late!*

I was having a crisis. I had prayed my whole life but never belonged to any church or religion. I intensely disliked church; and I felt more removed there than praying inside my own home or car. I prayed in several mosques in Africa, and had prayed in Buddhist temples in Thailand and Singapore. I was always smugly arrogant about not belonging to a religion. Why belong to one and betray the others? Now I was panicking, desperate to belong to something, and some ethereal, internal voice was chasing me. I knew it wouldn't leave me alone. I knew it was coming to call me out.

The voice persisted. "Listen Garrett. You worry you belong to none. Can't you see that by belonging to none you belong to all? You do not need to worry about this. You already believe in what is happening. I listen to you."

Just like that the voice disappeared, as did <u>all</u> of my anxiety. I felt safe all of a sudden, and quietly humbled.

I made it back to Dad's room. Tiff looked at me, blankly, like saying, 'Huh?'

All I could push out was, "Sorry, I just needed to use the bathroom, how's Dad?"

"He just had one of those jolts again, as you call them. He opened his eyes so wide, he looked so happy Garrett I swear it, like he had been somewhere else that was great. Then he just closed his eyes and fell asleep, or whatever this state he is in now."

"So more of the same?"

"Yeah, more of the same."

"Oh Tiff," I prepared myself to tell her of what just happened to me, about that voice that had appeared.

"Ya?"

"Oh its nothing, I'm just so sorry I had to use the bathroom so abruptly."

Tiff just smiled. "I'm freaked out too."

I never told her, or anyone else, what happened in that sunlit corridor, until writing these pages, just now.

"Oh and Garrett...? Like, I don't know what this means, it may not mean anything, but Dad said something before he fell asleep again."

"He did?"

"Yeah. He said, 'two days.'"

THE VISITS: DAY 4

Dad was still alive. Our family had all been to his room last evening, to say goodbye.... again. Dad seemed to rally last night and tried to tell us he was OK, that he was going to live in the shining city. He told Tiffanie's new husband Mike to take care of her, which reduced the room to tears. Then shortly thereafter Dad called the four grandchildren present to move in closer, so he could tell them he loved them. Of the four, only Shani and Lindsey, now both 18, could remember their Grampy as once being well and vibrant. The youngest two, Miah and Brityn, would only know their grandfather as someone with Alzheimer's who needed to be taken care of. Despite this, they were still emotionally close and bonded to him.

Dad had had nothing to eat for many days, and only a little water we could trickle down his throat from time to time. Dad's "electrical jolts" were continuing, but still not with any predictable regularity. We could tell when the visits were about to happen now, however, because Dad would begin tracing the arc over his head signaling the arrival of his "friends," who I likened to angels. Then he would get this giddy, childlike look on his face, as if he was an excited seven year old, getting ready to go to Disneyland.

Tiffanie and I were trying to assign meaning to Dad's "comment" the day before, when he told Tiffanie, "two days." Of course, we took this to mean two more days until his passing; wanting to believe that in two more days Dad would transition to the next world, to the "shining city." Perhaps, however, Dad's "two days" was just a random utterance coming from a brain that was no longer very organized due to Alzheimer's plaques, and meant nothing at all.

Today Dad said something different, however. He opened his eyes wide and said, "I want to go home, please... take me home."

"Ah...yes...ah... sure, Dad. We will get you home, and soon!" I spoke without understanding a damn word of what I had just said.

Tiff and I didn't know anymore what "home" meant for Dad. Over the past 7 years, "home" meant, and became, many things to him inside his Alzheimer's scrambled mind. Was it his last autonomous residence in Bay City? Was it his condo in East Lansing, where my family and I maintained a wraparound care structure for a couple years? Was home his childhood residence in Brooklyn? Or our family home in Los Angeles, where my sisters and I grew up? Maybe home was Traverse City, Michigan, where Dad had moved us all, after life in Los Angeles became too dangerous and stressful. Or maybe home was simply his memory care facility, where so many kind-hearted caregivers had become his new family.

Then again, was home this wonderful place in the afterlife, which he was being readied for? I so badly wanted to believe the latter, but the scientist that dominated my identity came to the most conservative conclusion. That was where I landed, despite what that voice had told me just yesterday.

I looked over at Tiff. "I think he means home at Claire Bridge, you know, he has so many familiar faces there. Maybe he's supposed to die there." My voice was calm and soothing for my sister, until I reached the word "die," and choked on it.

"Maybe we could try to figure out what he means?"

"Well, we could play Twenty Questions with him I guess!" I actually laughed at this. Something had to give.

"That's a great idea, Garrett! Hey Dad, do you mean the shining city is home, is that where you want to go?"

"Dad looked immediately frustrated. "No, no, no! The shining city, that's where...that's where I will be...well that will be my home! But right now I just want to go HOME."

"You mean back to Brooklyn, Dad?"

"No, no, you're not getting it. You may be the greatest but you don't know! Not home there, there's nobody there for me. My parents have moved, you know that!"

"Moved?"

"Yes moved! I told you that already! Don't you

remember anything? They live in the shining city, with my sister, you know, oh what's her name... Fr...Free..."

"Friedi."

"Yes, yes, that's it, that's it! She lives there too. See, you do know! And Connie lives there too, well not with my parents, but she lives in the shining city. I don't know where yet, but I saw her there and she...she... well she welcomed me."

"You mean Connie from Claire Bridge?"

Yes, yes, but not my girlfriend Connie, that's Constance, you know that one. She's my girlfriend. I mean my friend who was there, Connie!"

Tiff was confused. "There are two Connies that Dad is talking about, Garrett? I only know Constance, the one from Africa."

"Yes, he's talking about a resident named Connie. I am sure we've met her there, we just can't place her."

Tiff interjected, "Dad, you want to go to Claire Bridge, that's home right, to see Connie?"

"Yes, by God you've got it! Yes, that's home, that's the place. I want to go back there, right...right now! Only Connie won't be there. I told you, she's moved to the shining place."

"Remember, Tiff...he told us he saw Connie there a couple days ago...during one of his visits..."

Dad shook and quivered as another electrical "infusion and discharge" coursed through his body. "He's here, he's here!" Dad closed his eyes and was soon "asleep." He would later tell us that he was "flying" easily to the shining city, and that he knew the way.

Later that afternoon, Sue from Heartfelt Hospice, accompanied by Maggie, an Administrator from Claire Bridge, came to visit. We were told that they were trying to honor our wishes and let Dad pass in the hospital, rather than disrupt him and risk unnecessary pain or trauma. But they also gently told us if Dad hung on another day they would have to move him, that the insurance company would in essence refuse to pay and kick him out in favor of hospice. Our options were to transfer Dad to the hospital's in-house hospice care unit or bring him back via ambulance to Claire Bridge. The answer came to us easily.

I spoke for the two of us, this time formally, as his power of attorney. "He wants to go home. He told us earlier today, emphatically. We want to get him back to Claire Bridge. You guys have been so good to him there. He calls Claire Bridge home now."

Maggie answered with a kind smile, so typical of all the staff there. "Thank you. We will make arrangements and see if the hospital will let us wait until tomorrow morning. We will get his room ready at Claire Bridge. We have an electronic hospital bed that one of the families donated, we will get that set up for him and there will be no charge. His Medicare will cover nearly one hundred percent of this, so you don't have to worry much about additional cost."

Tiff then added, "He's been talking about his best friend, Connie. Not Constance, we know her, but is there someone else named Connie there, who's a resident?"

Maggie and Sue looked at each other, as if to say who is going to go first. This time it was Sue who took the lead. Though not Claire Bridge staff per se, she was assigned there by Heartfelt Hospice and knew most everyone in the building. "Ah, Garrett and Tiffanie, yes there was a Connie there, she was on hospice too…"

"Oh good, so Dad will be so glad to see her!" Tiffanie was hoping this would make Dad's transition to Claire Bridge easier.

"Well I don't know quite how to say this and I hope it won't upset you guys or your dad… but Connie died recently."

Tiff stared blankly.

"When?" I was almost afraid to ask.

"Two days ago."

THE VISITS: DAY 5

It was now Friday morning, over one week since Dad's condition had turned from something seemingly treatable to imminently terminal. Though Dad was tough as nails, the constant wheezing and weakening cough foretold the end. The hourglass of his life was still upside down, and running out of sand. All we could do was pray, and try to race against the clock to fulfill his last wish, to get him back home, to the familiar faces at the Claire Bridge memory care facility.

Although all the discharge papers were signed by 8:30 in the morning, and Heartfelt Hospice and Claire Bridge staff were ready to receive him, it seemed like an eternity before the ambulance workers showed up with a gurney at 11:30. When they arrived, Dad seemed to perk up a little, and he appeared excited and happy.

"You're here, you're here! You've come to take me home! I knew you'd come, isn't this great! I am going to go home today!" Dad tried to sit up and shake the ambulance workers' hands. It was classic Dad, always bent on showing gratitude and respect, no matter what one's lot or position in life was.

With great precision and sensitivity, Dad was lifted from his bed and strapped in to the gurney. If he had any fear or apprehension he didn't show it. He looked secure, and totally at peace.

Dad looked over at Tiff and me. "Well, anchors aweigh!" Dad used his old Navy term for the first time in years, and saluted us with a seriousness that I didn't expect. I saluted back and smiled.

"So I will see you back home. "

"Yes Dad, you will. We will get there about the same time you will."

The ambulance worker then gave us an unexpected gift. "So, who wants to ride back in the ambulance with your dad. We can take one."

Tiff said "you go with him, Garrett. I'll follow you guys over. We can come get your minivan later."

Dad lit up like an old pinball machine. "You mean you can come with me?"

"Yes, Dad. Anchors aweigh!" And with that send-off line, Dad was wheeled to the service elevator, which released us in some sort of sally port where the ambulance was waiting. The driver greeted us, Dad saluted again, and we were off. Thirty minutes later, we arrived at Claire Bridge, where a group of caregiver staff was waiting for us at the door with an automated hospital bed. Dad was delicately lifted from the gurney and secured in the bed, propped up at a 45-degree angle. He would stay at that angle for the duration of his time in that bed.

Inside, we were greeted by a host of staff, lined up like a welcoming ceremony at the United Nations. The caregivers at Claire Bridge were from all over the world, many of them medical or nursing students from Michigan State University. Many were immigrants from Africa, who brought their compassionate respect for the elderly with them. I was known to the African staff not as "Garrett," but as "Walter's son." That precious identity was never lost at Claire Bridge.

Dad was wheeled past the welcoming committee and saluted several times. His eyes welled up with tears and he said, "they are all here, for me, I am finally back home." Each and every staff we passed touched or kissed him gently on the cheek or forehead. The respect for this now frail man, once called The Champion, was mind-blowing. I wondered if they treated everyone placed at Claire Bridge with this level of respect and compassion.

Dad was dying, the welcoming was also a farewell, but he seemed so happy. When I later told my mother about this, she told me something I had already suspected, but was now confirmed. "Your father has been searching for unconditional love his whole life. He didn't feel it as a child, nor as an adult. I think he is finally learning this lesson, before he dies. It is a gift for him."

Dad's entourage followed us into his room, which they had immaculately prepared for him. Many people from many cultures prayed over his bedside, each in their own unique way. I made a list on my now 20 plus page notepad from where every one of the greeters came: Malawi,

Zambia, Kenya, Tanzania, Botswana, Nigeria, Cote D'Ivoire, Senegal. I was struck by how different the cultural rituals were; yet their prayers all seemed the same. One caregiver's prayer seemed all-inclusive of what everyone was saying. "Have a good journey, dear one, to your new home in the afterlife."

The welcoming farewell continued for several hours, ebbing and flowing with the Claire Bridge shift change. As staff left for the weekend, I was sure they were certain that when they returned for their next shift that they would find the room empty.

By 4 PM, Dad had drifted off to sleep. At 5 something PM, he jolted, tried to rise up in his bed, and exclaimed, "They are here, they are here!" Tracing an arc over his head, he counted, "one, two, three, four, five, five, five, now six!" He said with a force I didn't know he had remaining in him: "They came back to go home with me!"

THE VISITS: DAY 6

I had spent the night with Dad but I didn't sleep at all. My mind raced and I could never settle down. One would think that I would have been filled with inner conflict, anxiety, or fear at my dad's impending death, as he wheezed and coughed nearby. But I did not feel any such angst. In fact, I was rather greedy in a way. I was beginning to believe this new plane of existence Dad was going to was not a hallucination elicited by the death process; this seemed too real. I was now waiting for Dad's passage and wanted to see it. On this night, for a time, it wasn't about Dad anymore. Having accepted and now anticipating his passage, I wanted to see it. I wanted to be there when the miracle occurred.

Around midnight, I felt a presence about me, just like the "visit" I received in the hospital three days before. As before, it was a voice from within my head that seemed external. I wanted to know why it had come for me and not my dad. Though I could "hear" the voice, it spoke without any sound. "More miracles are coming. You will be able to see, because you are not handicapped by religious structure." I could feel the energy leave the room, and with that, the presence ended. Feeling as if no one would believe me, I would tell no one of this experience.

Dad had numerous "jolting episodes" the remainder of that night. At one point he tried to get out of bed and I had to call in an aide, Ruby, to help settle him back down. We tried to give him some morphine to get him to sleep, but he spit it out several times and we just gave up. Although wheezing, I also noticed Dad no longer looked sweaty and feverish, and he no longer felt hot. It appeared that after a week of fluctuating fevers, the fever was gone. I took this as a sign his body had given up the fight.

His eyes opening wide, Dad tried to sit up in his bed around 5 AM. Now accustomed to this occurrence, I simply waited for the arrival of the entourage of angels. It was the only description that made sense to me. And right on cue, Dad used his fore finger to trace the arc over his head. I was beginning to realize that the end of this life was no longer bad, or something to be afraid of. I calmly asked Dad if it was his time to die, and if so, he could just stay in heaven this time without coming back.

He reached over and patted me on the head, "My boy." It was the first time Dad seemed to know I was his son in over a year. Choking, I said, "I love you Dad," truly thinking that would be the last thing I was going to ever say to him. I thought the miracle was coming. He was going to pass over to the next realm, to the place he called, "the shining city."

Dad's body thrust upward as he opened his eyes, with an animated look on his face. "He's coming now!"

Then Dad said it again, just like last week in the hospital.

"Two days." This time he pointed to his U.S. Navy watch. Dad had lost the ability to estimate time, including how to tell time, over three years before, but he steadfastly refused to let anyone take that watch off. It seemed to help secure the last vestiges of an identity for him.

"You know how much time you have, Dad?"

"Yeah, I do know." Dad smiled a smug little smile.

"How much time do you have?"

Dad now smiled brightly. If he had any discomfort, emotional or physical, you would never know it.

"Yes, I know."

"How much time do you have? " I persisted. Dad shook his finger in the air as if to say, 'not so fast, that's for me to know, not you.'

When he closed his eyes, presumably to be with "the kind man with the grey beard," Dad said, "He's so beautiful."

He also said, "I look so good where I am going," as if to imply that he was again youthful and healthy on his journey.

Dad was asleep or unconscious for the next six hours. I stayed in his room, unwilling to leave. I wanted to see the miracle. I envisioned it the only way I guess I could. An angel would appear, the room would light up, and Dad would somehow ascend to the "shiny city."

None of that happened.

THE VISITS: DAY 7

It was now Saturday, one full day after Dad left the hospital, had returned to Claire Bridge, and received his "ceremony of goodbyes" from the staff. Dad had not really eaten anything in well over a week, but he was still willing to take in electrolyte-bolstered fluids occasionally. I sat at the foot of my adopted bed this morning, wondering if Dad would ever awaken again. He had been "out" for some hours now, and the jolting had stopped for a longer period than usual. I wondered if he was at the shining city and was getting ready to stay this time. I thought for a fleeting moment that I should wake him to get some more fluids in him. I answered my thought out loud, "what does it matter?"

One of the senior caregivers, Constance, came into the room. She was the other "Connie" in Dad's world. Constance was a beautiful thirty-something woman from Malawi, with a heart of gold and an all-enveloping serene presence that followed her. Dad called her "my girlfriend," and often held her hand.

Constance sat down on the bed next to me. I had known her for two years now, and usually the two of us together were never at a loss for words. However, we now seemed content to just sit in the quiet. She smiled, and I smiled.

Finally after many minutes, Constance spoke. She almost never talked for the sake of talking, always trying to impart something meaningful. This time was no exception.

"You know, Garrett, Walter's proud son, when I came to the United States my father was very sick and he was dying. I didn't want to leave him. I just wanted to be by his side when it was his time to make his journey. I was very close to my father, just like you. But my sister and I had won the lottery selection to come to America, and my father knew we just had this one chance. So he told me to go, to take care of our family, and to not worry about him. We had to respect his wishes. The day after we arrived in America, he died."

I looked into Constance's tear soaked eyes.

"For a long time I was so angry at him, and at myself for leaving him. Now I have come to realize that people often know when it is their time, and a certain peace comes over them."

Constance was now crying. I had never seen her in any light except composed. She was not composed now.

"Garrett, I know you want to remain by his side, when he goes, but, you see... I have come to realize, I have seen this so many times here on hospice, that most people want to be alone when they die. I am not sure why, maybe its because they sense you don't want them to go and they are unable to let go because of this."

Constance put her hand on my knee. "You may have to leave your father's side for him to be at peace with his journey."

"And besides, your family hasn't seen very much of you. You are always here. I think they want their father and husband back home. Your father... well he is in God's hands, he has been trying to tell all of us for some time now."

With that, Constance got up, walked over to my father, kissed him on his forehead, made some sort of prayer gesture, and left the room with tears streaming.

I went home that night for the first time in days.

THE VISITS: DAY 8

It was Sunday, nearly a week now since Dad had been taken off all life supports and antibiotics. I had gone home last night at Constance's impassioned directive, to get some rest, be with my family, and give Dad his space. Two of the three tasks had been accomplished, but I failed on the rest part. I just couldn't seem to sleep any more. I got up early this day, and raced down to Claire Bridge, not knowing what to expect. I didn't get any emergency calls over night. On the way there, I called Tiffanie, and she said she hadn't received any calls overnight either.

For some reason, probably now a heavenly one, Dad continued to hang on. It was now becoming abundantly clear that Dad was going to die on his own time, and on his own terms. I still didn't know what the "two days" meant, and thought maybe it was just something random that Dad said as it crossed his certainly disorganized mind. Nothing really felt random, however. Dad's "experiences" seemed so consistent, almost predictable, and the content always the same. If these were simply hallucinations, how could they be so organized, so lucid, so consistent?

In his room, Dad seemed to be asleep, though I never knew if he was really asleep, unconscious, or having one of his out of body experiences and had vacated his body. I was becoming increasingly comfortable with the notion that Dad really was leaving his body, and somehow "flying" or otherwise being transported to another "place," or realm of existence.

A pleasant daydream of Dad and me jumping the big waves at Zuma Beach, our favorite Los Angeles retreat, came over me as I held my father's hand. I started to fall asleep, but startled when one of the familiar care staff entered the room. It was Sarif.

Sarif was an extremely pleasant, and very humble, middle-aged man who had come to the United States from Tanzania, East Africa. He was studying computer science in college but somehow still managed to work full time at Claire Bridge. He was nearly a daily fixture at the facility and I saw him more than any other staff. I knew he was deeply spiritual and a practicing Muslim.

'Oh, hi Sarif!"

"Walter's son, I am so sorry to have awakened you. I am sure you are exhausted. I am sorry."

"Dad is still with us."

Sarif smiled. "I see that. We must try not to guess what God already knows."

Sarif may or may not have known how absolutely profound that statement was for me. For me, his message had about five meanings all at once: Let go, let him be, he is on God's timetable not ours, what do the doctors really know, and maybe Dad has a purpose for hanging around. All the messages reached me at the same instant.

Then I said something that to this day I cannot understand, but Sarif took in stride, "Would you help serve as our spiritual advisor?"

"Yes, yes I will do that, Walter's son. Would you please stand up? We will pray. But since Walter is still with us, we must pray that he will get better, not guess that it is his time to leave. That choice is not ours."

"Yes, I understand."

"In my culture, we lift our hands up, like this, with your palms upward. You know this already from your visits to my continent. Now we pray. We pray for him to get better."

I lowered my head as Sarif said a long prayer in mixed Swahili and Arabic. I couldn't make out any of the meaning except "Amin" at the end, the equivalent of "Amen," and "Inshallah," or "God-willing."

Sarif said, "So there, it is done. I will be back to change him in a few minutes… you can continue to say some more prayers if you want." And with that, he walked out.

Two minutes later Dad stirred, opened his eyes, and fought hard to re-position himself. It looked like he was trying to sit up and turn himself to get out of bed!

Startled, I pulled the emergency cord attached to Dad's bed. Within seconds, in raced Ruby, from Botswana, and Sarif. Dad had managed to position himself perpendicular in the bed, and now it looked like he was trying to push himself up to swing his legs over the side of the bed.

Ruby and Sarif tried to assist him, one on each side. I expected them to try to lay him back down. Instead, it appeared they were trying to get him to stand up!

"I don't think we should…" I said this so feebly no one heard it. Besides, I didn't want to complete the sentence anyway. I wanted Dad to stand up, just like they did. I wanted Dad to be The Champion one more time.

Constance entered the room. Almost as if scripted out, she positioned herself in front of Dad, with Ruby and Sarif at each arm. Constance seemed to evoke the Holy Spirit. "Come on Doctor Turke, you are going to walk today!"

With that directive Dad was pulled to his feet. One, two, three, four steps…with assistance. Amazingly, his balance was good, and he made no effort to try to sit back down.

"Come on Doctor Turke, U.S. Navy Man, Garrett's proud father. Today you will remember what it is like to walk again." Constance took off an ooccupational therapy gait belt strapped to her, put it around Dad's waist, pulled it taut to help keep him balanced, and said, "lets get him to the cafeteria."

Dad walked, arm in arm with his three-person support

team, out the door…into the lobby…and into the cafeteria! There we helped him sit down. A bowl of pureed oatmeal was brought to us and I started spoon-feeding my dad. He ate easily. I was completely stunned.

"God is good, yes, Walter's son?" Sarif smiled.

I smiled back. "Amin. God is good."

I remembered the voice that came to me two days before. I thought to myself, *Well, there's the first miracle.* And then I "heard" back, "More will come."

I could hear the electronic door chime in the background, signaling the arrival of visitors to the facility. Soon I could hear Stacy, 11 year-old Miah, and 18 year-old Shani. I ran out to greet them. As I expected, they were walking in the opposite direction, toward Dad's room.

"Hey you guys, come this way!"

Stacy looked a little puzzled. "Oh…sorry. Walter's still asleep."

"Ah…no….he's not…"

"Changing him?"

"Ah, no…"

"Huh?"

"Where's Grampy, Dad?" Miah was always direct and to the point.

"Ah…he's eating breakfast in the dining room."

I will never, ever, forget the look of utter astonishment on their faces.

"Come on, you can help feed him!"

Stacy, Miah and Shani sat down next to Dad in absolute amazement. Shani started spooning oatmeal into his mouth.

Miah turned to Stacy and simply said, as if it was an obvious fact, "Mom. He's not sick any more."

I said, "I've gotta go back to his room for a minute, could you finish feeding him?"

I went back to Dad's room, and checked Dad's bed. I guess I wanted to confirm what I was experiencing was real, and not some sort of dream.

A couple minutes later, Miah ran into the room so fast she was out of breath. Excited, she couldn't get the words out quick enough. "Dad, come quick come quick! Grampy is trying to write something!"

I raced back to find a group of people surrounding Dad, with Stacy positioning herself to try to assist him with a pad and a pen. Dad was attempting to scrawl something. "One, two, three, three, four, four, four…five…" the scrawling was unrecognizable, but we knew what he was trying to write. It was that number sequence again.

THE VISITS: DAY 9

Dad seemed exhausted after his walk to breakfast the day before. He slept hard all night, but his breathing was still marked by wheezing and an occasional weak, unproductive cough. His fevers, however, were gone. Although I prayed mightily for Dad to remain with us, I still had the sense his time was coming soon. Maybe what we saw yesterday was that brief moment of lucidity and energy that sometimes occurs before someone passes away. The hospital staff the week before seemed well versed in that occurrence, and said it was a common phenomenon among among terminally ill patients. I thought that maybe that was part of the life cycle, a gift of sorts for people to say goodbye or put things right in their life.

But that thought was met with a philosophical counter-thought, which seemed to be part of my scientific upbringing: *well that might be true for some, but what about all the people that die abruptly in accidents, cruelly at the hands of others, or painfully in the wake of some ravaging disease?* The model of a peaceful, natural transition to death seemed reserved for a tiny, maybe precious minority.

Dad interrupted my little internal debate. He suddenly stirred in his bed, his eyes grew large and wide, and he exclaimed excitedly, "He's here!" He then tracked "him" in the usual arc over his head, pointing as if everyone could see "him."

"Is he talking to you Dad? Is it OK to ask him questions?"

"He says I am ready to go to the shining city on my own now, I know the way. Oh! Oh! I am going there soon! I am going there! I am so at peace! Oh there is singing and dancing there, I get to play, too." Dad was laughing now. It wasn't the deep, guttural laugh I remembered all the way back to my childhood. This was a light, giddy, excited laugh, like an innocent little kid.

"Today I get to be a guide!"

"What? A guide? For whom? Me, Dad? Are you going to be a guide for me?"

"No, no, no. Of course not, not for you!" Dad smirked like the answer was so obvious.

"I get to help him! I get to show others the way…its, its, like my…" Dad drifted away before finishing, his face all aglow.

"Its like what? Your new….what?"

"Its my new, how do you say it, like my new…"

"Job?"

"Yes! That's not it but close enough. You get it! Its like my new job!"

"Yes, yes. I have another man to take today. We

78

will fly!"

"Who is this other man?"

"That doesn't matter, I just need to help him get there."

And with that, Dad was gone. The jolting stopped, and he was either asleep, or had just vacated his body.

An hour later, another jolt ran up and down his body. Dad opened his eyes. "I saw my mother and my father again today."

I then decided to try a little test. I thought of Dad's hero, the British actor Richard Harris, who had passed away a few years before.

"Could you find Richard Harris there?"

"Easily."

"Jackie Robinson?" Another one of Dad's heroes.

"Yes of course."

"And Glenn Miller?"

"Quit joking with me, you know the answer."

Dad then looked right over at me. "I can find anyone who is dead."

A transcending shiver now ran up and down my spine. I moved closer to him, as he seemed to be trying to reach out for me.

Dad touched my arm and held it, weakly trying to pull it closer.

"I think, " Dad started stroking my face, like I remember he used to when I was a boy.

"I think...you. I think you.... are.... perfect."

Dad then started crying. I started crying. I was convinced this was the moment he was finally saying good-bye, and that he was about to leave this earth.

"I am at peace. No fears, no worries, and no......"

Dad searched for what I was anticipating as his last word on this earth. The silence was just a few seconds, but it seemed to just hang in the air for an eternity.

"And no cash!" Dad let out a powerful laugh, the deep, guttural kind that had always defined him. "Yeah I don't need any cash there!" And with that, Dad drifted off again.

The rest of the day was quiet. I remained vigilantly by Dad's side, waiting for his ethereal "ascension."

Once again, Dad did not die on cue.

THE VISITS: DAY 10

It was now a week since Dad left the hospital, and over 10 days since we had been told that Dad would have no more than a couple more days to live. He hadn't had any fever in several days. He still wheezed and rattled when breathing, but he no longer seemed to breathe erratically and sometimes struggle for air. Dad was eating solid food again, though everything we fed him we had to puree and feed him in very small amounts. We couldn't risk him aspirating any food with his lungs in such bad shape.

Today we faced another problem, however, that could threaten Dad's life. He was developing bedsores from being confined to his bed for so long. Despite being turned and repositioned by nursing and care staff, he needed to get up more and if possible, walk a little each day with assistance. Aside from the one miracle day when he walked to the cafeteria, Dad seemed reluctant to try to get out of bed even when prompted. This morning was no exception.

"Come on Doctor, it's a beautiful morning, lets get up and get changed and go out for breakfast!" Constance's effervescence filled the room. She was a force to be reckoned with, a combination of drill sergeant and God's right hand angel.

Dad pushed her arm away. This was not unusual for him with other staff, but I had never seen this when he was with Constance.

"Doctor Turke, it's a fine day out and you are not going to miss it. After everything you have been through and overcome by the grace of God, I am not going to let you succumb to bedsores. Lets get going now! Come, come, its time to rise and shine!"

"I can't do this anymore." Dad actually seemed irritated.

He spoke again. "I can't do this anymore. I want to go back to the shining city. I can walk there, and I am 30."

Constance called for assistance and soon Deb, the shift manager, came in to help. They pulled Dad to his feet and steadied him, then began the 5-minute routine of changing his underwear and the "medical scrubs" he liked to wear when in bed. Dad looked very unsteady on his feet and was breathing hard as they did this. He seemed out of breath and exhausted from the experience.

Once changed, Constance and Deb each grabbed an arm and tried to get him to walk a little. Dad would have none of that today. "I told you, I can't do this anymore." Dad's voice had an angry forcefulness to it I hadn't heard in a very long time. He pushed back on Constance and Deb's efforts to escort him just a few steps. Defeated, we all then helped Dad just sit down on the edge of the bed, propped up with some pillows. That would be the best we could do today.

I began to realize that Dad's recent "recovery" might have been more illusion than fact. He had only had that one good day when he walked to the cafeteria. Now he was back to being bedridden and weak, out of breath from simply being changed, and talking as if he was all done with any kind of mobility. Was he telling us to simply let go, and let him continue his physical erosion in exchange for his journey to the next life?

I broke up the loud silence inside my head by putting on the film version of Camelot soundtrack, with Richard Harris, Vanessa Redgrave, and Franco Nero. When Dad's favorite song came on, much to my amazement, he started singing.

"If ever I should leave you...

It wouldn't be in summer...

Seeing you in summer, I never would go... "

I was stunned by the fluidity of Dad's singing. It wasn't that long ago that Dad had lost virtually all his language. It had come back while in the hospital. But this was astonishing.

"You know I saw him?"

"Saw who, Dad, your guide, or one of the angels?"

"No, no, no!" Dad pointed to the boom box.

"Him!"

"Franco Nero?" He had played Lancelot, the character singing this song.

"No, no. Not him. You know, the other one, the big guy!"

"Richard Harris?"

Dad grew excited; his eyes opening wide. "Yes, that's the one! I saw him over there!"

"I can see anyone I want!"

"Anyone?"

Dad laughed. "Well, they have to be dead."

Constance entered the room to see how we were doing.

Dad saluted her, and Constance gave an exaggerated salute back. "Doctor Turke, sir! U.S. Navy man. Freedom fighter! Sir!"

"Dad's been singing, Constance! But I don't think he's gonna walk today."

"Well then, I will talk to Deb and we will probably have to put some booties and ankle doughnuts on to keep his circulation maximized."

"Yeah, I figured that would be next. Hey, what time do you have? I forgot my watch."

Before Constance could answer, Dad did.

"It's time to go back to the beginning. They are here."

Dad chuckled to himself, leaned back against the pillows propping him up, closed his eyes, jolted, and left for another "visit."

THE VISITS: DAY 11

I awoke in Dad's room on my makeshift cot across from Dad's bed. I had a dream that Dad had died.

Tiffanie, Brityn, Lindsey, Stacy, Miah and Shani all came into the room. They brought flowers for Dad's bedside, some bagels for us, and some pureed "mush" from the kitchen, which I was informed was oatmeal mixed with fruit cocktail and Ensure. They were here to see Dad, who was now being called "The Miracle Man" by staff and my family alike. All of the doctors and most of the adults in our family thought Dad's hanging on was a sign of his emotional and physical strength, and maybe his unwillingness to let go of the people he loved. They all still thought, though, Dad was going to succumb to his lung infection, and that his death was inevitable.

Our kids did not believe a word of it. They kept saying their Grampy was getting better, and wondered how we could miss the obvious and miraculous recovery he was making.

"He looks great, his coloring is back to being that ruddy red complexion again," Lindsey said with a burst of enthusiasm so loud it caused Dad to stir from his sleep, or whatever he was doing with his eyes closed.

"Yeah and his breathing seems much better," added Shani.

Stacy offered the first adult opinion. "There is no doubt that Grampy hanging on for so long is a miracle, but I think we have to be cautious not to get our hopes up. Remember, his lungs are still filled with infection and he is so weak any progression in this infection will be hard for him to fight off."

The kids didn't want to hear it, but they tried to be as diplomatic as they could. Shani spoke for them. "Yes, Mom, you're probably right, but Mom, he doesn't look like he did last week. I think we have to get him moving, so he can get back into shape again."

10 year-old Brityn chipped in. "Uncle Garrett, what's his fever been like?"

"He hasn't had a fever, any fever, in several days now. Once it was 99.8, but that's still normal..."

"Or a mild fever," Tiffanie, like Stacy, didn't want to get our hopes up.

"And they are still giving him Tylenol, so he could have a low grade fever and the meds are keeping it down..." Stacy and Tiff were on the same page.

Lindsey and Brityn went over to Dad's bedside, as he appeared to be slowly waking up. Lindsey started stroking Dad's stubbled face. "Hey! Grampy! Open your eyes, look who's here to see you!"

Dad's eyes opened, but he looked like he was still in another realm.

Tiff moved in with a chair, and readied herself with a spoon to start feeding Dad the pureed oatmeal.

Dad reached out into open air. "The angels are right here. Right now. Can you see them?" Dad pointed to several places about the room. He then scanned over the group of us, nodding as his glance caught each person.

"Thank you all for coming to see me." Dad was true to form. He was always so appreciative, all the time, to nearly anyone who showed just a hint of kindness toward him.

"They are all here." Dad made a sweeping motion with his arm and hand, as if to include the entire room. We thought he might be talking about us, but he wasn't.

"All my angels are here, waiting for me....may they keep us safe, and sound."

Then Dad, pointed at me, and made a motion to come closer. "They are all around you now." Dad patted me on the top of my head and started stroking my hair.

An all-enveloping "shiver" ran up and down my spine. I was hoping no one would see it.

"Yes, Dad, I can feel them. They are all here, to keep us safe and sound."

"Now you got it, now you understand! Yes, that's their job...to keep us safe..."

"...and sound!" Brityn finished his sentence, and Dad laughed.

Lindsey looked at all of us. "You guys, I'm pretty sure now that Grampy is not going to die. We must be like the staff here, we must insist that he recovers."

The kids all agreed enthusiastically. They began to take turns feeding Dad.

The adults just quietly smiled. We were the cautious ones. Our innocence had already been trampled upon by life.

Later that evening, I returned home to wash my clothes and to take a shower. I decided to take advantage of the government's Family Leave Act to buy more time with my dad. I still wanted to be there when he passed, but I was now being pulled in many other directions. What if the kids were right? Was Dad truly recovering? Unjaded, maybe they could see the obvious when we adults could not. Maybe Grampy was indeed getting better.

While taking a shower I decided to stay with my kids and my wife for the night. At first, I had to fight off some guilt in order to make that decision. It was then that I felt that presence again, the energy or spirit that had visited me before. I waited for "it" to speak to me again, and it did. "Do not be concerned with earthly matters. This will be all right."

The presence then left as quickly as it appeared. I raced out of the shower and wrote down the words that were given to me.

I played basketball with my kids and Stacy made a nice dinner. I slept hard that night. I didn't know it at the time, but I would need that deep, resting sleep in order to handle what would be coming tomorrow.

THE VISITS: DAY 12

It was raining outside, but a bright day was unfolding inside Dad's room. Dad woke up smiling from the start, and I watched him perform a sort of ritual that was becoming more commonplace for him. He would wake up, and seemingly start studying his hands, then his face by touching it, over and over. It took me awhile to finally appreciate the significance of this and what he was doing. Well, what I thought he was doing. It was just my interpretation of course, but I thought Dad was checking to see if he was "in" his body, and not in some other spiritual realm, or dimension.

"I am so relieved. Very relieved." Dad seemed to be talking to himself. Then he started crying. Dad removed his stroking fingers from his face and pointed with his index finger above his head, moving his finger from side to side in an arc. I was accustomed to this behavior already of course; he was tracing the presence of what I could only call angels, over his head.

Constance entered the room to check Dad's status for the morning. Dad outstretched both his hands, caught her, and pulled her in tight. "You are so, so, you are so, so, so…beautiful!"

"Why thank you, Doctor Turke, U.S. Navy man!" Dad patted Constance on the back, pulling her in with his signature bear hug.

I put on some music on his small boom box stereo next to his bed. Dad's favorite album, post dementia, was already loaded in the CD tray. Just about everyone in my family, and in this building, knew that Dad's favorite artist was Don McLean, who scored a big hit, decades before, with American Pie. Yet that was not Dad's favorite song on the album. Dad absolutely loved "Vincent," about the self-tortured artist Vincent van Gogh. Dad had always identified with his pain, captured beautifully by McLean's words and haunting melody.

Dad moved his hands in a rhythmic, maestro kind of way. "This is the best, the very best..." I watched a tear roll down Dad's eyes. He was thoroughly "present," conscious, and in the moment. His Alzheimer's had robbed him of his "connectedness" for years now, and I thought that connectedness was forever lost. Well not today, apparently.

Then something happened that I didn't expect. Dad spoke with a confidence I had not seen in a very long time. He sounded serious, intent and prophetic. I listened carefully, and quickly grabbed my pad to write his words down.

"Today there will be a reprieve. Another miracle is still to come. The angels, well there's more and more now. Soon it will be time to rise up! Yes, and I will look forward to it, yes!" And with that said, Dad started crying again.

"Dad, don't be sad." I actually had no idea of what he was actually feeling. I had not heard him speak with such confidence and fluidity in years.

"I am sad."

"Why, Dad?"

Dad moved his mouth as if to form words, yet no words emerged.

"Why Dad, because you have to say goodbye?"

"Yes." Dad reached out and caught me, pulling me in for another bear hug.

"Don't be sad, Dad. We want you to rise up! Soon! Not to be stuck in this bed, in this tired old body."

Then I added something without even thinking. I say, "without even thinking," probably because when I said this I couldn't believe what I was beginning to grasp.

"Dad, we will need you to be one of our guides when our time comes." I couldn't believe I said it. I couldn't believe I seemingly meant it. My more conscious, Western trained mind still wafted in disbelief.

"So, Dad, you have to die first!" I said it like a joke, but I actually meant it. Dad laughed, though I thought he might just have been mimicking my laugher.

I tried to undo my joke. "Yes, Dad, it's a miracle going on, just for you!"

Dad started crying again, but in a seemingly happy, I-am-touched, sort of way. But then he abruptly started crying hard, which had a totally different feel and connotation.

"Did the angel walk for you last Wednesday?" I was trying to tap Dad's apparent power, in the moment.

"Yes, yes! How did you know?" Dad was now crying rather forcefully. I worried that I had triggered this, and had no idea where it would lead. But his crying soon stopped, followed by a few minutes of quiet, after which Dad fell "asleep."

An hour later, Tiff came in to cover the next "shift," and give me a break. Dad was still asleep, but before I left, he startled himself awake.

"I woke up…" Dad chuckled to himself.

He turned his head, apparently looking outside at the rainy but warm late summer's day.

"What are you looking at, Dad?"

"The angel."

Dad looked at his watch, smiling a big, wide-eyed smile.

"Do you know how much time you have, Dad?"

"Yes."

"How much time is that?"

"Not much."

THE VISITS: DAY 13

It sounded so awkward, but I was still searching for Dad's death. There was no other way to really explain how I felt. Why was this taking so long?

Part of me wanted Dad to hang on, he was still my dad and he was The Champion, at least for me. I wanted to still have him there to talk to, to depend on, and to believe he was still teaching and guiding me as my father. As I looked upon him in his bed, all I really saw now was the nearly empty husk of a man who used to be so strong and robust. It was selfish of me to have him hang on. There was a better world waiting for him, beckoning him each day to pass over from this life into the next. Was he hanging on just because of my selfish wants? Was this the reason he couldn't let go?

I put Dad's music on, although very softly, and soon Don McLean's Vincent came on.

Dad smiled, and tried to sing a little, "Starry, Starry night..."

"Hey Dad, I've been thinking. I know you want to join the others in Heaven." That was the first time I used that word, instead of "the shining city."

"Well... I ah wanted you to know... ah... Dad, I wanted you to know that....that..." I was choking on my words. The thought that I wanted to convey could be expressed cleanly in my mind, but I couldn't get the words out.

"Well, you know, I love you very much....and... ah, you know, we had a great life together, all those adventures we had..."

I struggled to continue. Dad was looking at me, but with a passive countenance. I couldn't really tell if he was paying attention to me at all.

"Yeah, Dad, there were some tough times too, we all made our mistakes, but that's all water under the bridge. Everyone makes mistakes, we are all so flawed...all that is forgiven, you know..."

I noticed a single tear trickle down from Dad's right eye. Maybe he was following me, after all.

"Well I just want you to know Dad...its OK for you to go now, we know you are going to a wonderful place, so I wanted to say goodbye now, so...ah....so... ah when you do leave we don't have to be sad, we can just be happy for you, I mean, when your time comes."

Dad didn't say a word, but the single tear had become many tears, from both eyes. Finally, Dad just smiled faintly, the tears streaming down his face and shimmering in the bedside lamplight. He tapped the face of his watch a couple times, which I interpreted as signifying that time was running out.

Dad spoke…"Jeeez…. Jesus!" I couldn't tell if Dad was trying to impart something, whether he was summoning, or whether he was just using Jesus as an exclamation point like people do.

Dad's eyes grew huge and he tried to sit up all the way, arching his back to move. For a minute, I thought he was getting ready to try to get out of bed. He didn't look as if he was "present" anymore however, appearing as if he was just staring right through things.

Dad then said something I will never forget. He said it as clear as a bell, and with some authority.

"Walter Turke, I am here!"

For a moment I thought it wasn't Dad who was speaking. Then my words followed suit.

"Is God in you right now?"

"Yes, yes."

"Are you in heaven, are you walking?"

"Yes. Its…its another miracle." Dad pointed to thin air, and then traced the arc over his head with his forefinger. "He's right, he's right over there now."

"He's no longer inside of you?"

"No, he's over there now," Dad pointed to an empty corner of the room.

Dad seemed to be in the two worlds simultaneously.

"I saw him and I told him I loved him. He has been dead a long time. He's about 30 years old, too!"

Dad moved his arms gracefully to the music, as if he was conducting an orchestra. I was becoming inpatient. I wanted more answers.

"Who did you see? Who did you tell that you loved him?"

"Him. You know. Him."

"Jesus? God? The man with the grey beard? An angel?"

Dad shook his head and smiled, as if the answer was obvious.

"There's three new ones."

"What?"

"Yes. I know them all!"

"Who are they?"

Dad just shook his head again, with that playful, almost impish smile that seemed to say, I'm not telling.

"We need to help them." Dad sounded like he was giving a directive.

"We?" The thought of me joining Dad on a mission in the other world sent a chill rippling up and down my spine.

"Yes, we. You need to help them too." Dad's speech was fluid. His Alzheimer's disease was being lifted again.

"How?"

"You help them by just believing what they say. Its easy."

"You mean you want me to just believe them?"

"And now, my boy, you've got it."

A tear rolled down my eye.

"Just do what they say. Its easy!" Dad looked joyful.

"Will they tell me what to do?"

"Yes, when it's your day." Dad pointed down at his watch again; lightly tapping it.

I sighed with some relief. Dad was talking about the future. "I understand. You just want me to believe…"

"Yes, yes! That's it!"

"Can I ask him a question?"

I still didn't really know who "him" was. Were they angels? God? Dad's guide? Jesus? Maybe they were all manifestations of the same thing.

"Yes," Dad replied softly.

I asked a very brave question. "Dad, why are you still

here? You were supposed to have died many days ago."

Dad paused for a while, like he went somewhere to retrieve an answer, and then spoke some more.

"I'm supposed to..." Dad's speech became garbled. I couldn't make it out.

"Dad I couldn't quite hear you, I am supposed to..."

"I am supposed to come down here and..."

Just then the Clare Bridge nurse, Angela, burst into the room, aglow with positive energy. Dad was distracted by Angela, and never finished his sentence.

"I just wanted to see how your dad was doing, Garrett." She started re-positioning him, beginning the now accustomed, multiple times per day ritual to lessen his risk for bedsores.

Angela was still speaking to me but was looking right at Dad. "I am just amazed by this man. He truly is the miracle man, as the staff says."

I was going to answer but Dad interrupted me, to talk to Angela.

"I am about 30 over there and I walk. Just like him!"

That was all Dad would share for the rest of the day. He drifted in and out of "sleep," with the occasional, accompanying jolt. To say I was disappointed would be the understatement of the century.

I never got an answer that day, or ever again, for what Dad was supposed to still do on this earth.

THE VISITS: DAY 14

Day fourteen of Dad's ethereal visits would be his most quiet. I had gone home the night before, now confident that Dad was not going to pass away on some timetable, and not as some sort of spectacle for me to see. Dad's breathing still wheezed and rattled as if he was still fighting pneumonia. It no longer had a desperate sound to it, however, and it no longer sounded like it was worsening. He had not had a fever in over a week.

I had a fleeting thought that maybe Dad was no longer dying. I quickly dismissed that. After all, the doctors were so certain and convincing that Dad's recovery would be impossible. A nearly 80 year-old man simply does not recover from two collapsed, pus filled lungs.

Dad spent "Day 14" absent. He was "asleep," or perhaps a better word would be "traveling," the entire day while I was there. I could not arouse him nor really wanted to. He jolted on and off periodically, the energy or electrical current traversing the length of his entire body like it always did. Dad was certainly off somewhere. I had a thought enter my head, which I then said out loud. "I'm going to prepare a place..."

I imagined Dad to be in the shining city, walking confidently as a thirty year old man. I pictured him smiling, walking past a group of happy people in a circle, who were singing and dancing. I envisioned him walking toward some vague but shining house, where his mom and dad, and his sister greeted him. In my mind, they were all young there, too.

I chuckled out loud as I returned from my little imaginary trip. "That's crazy, Garrett, that's just crazy." I was talking to myself.

Dad lay quiet the whole day. I stayed with him and read. Dad never opened his eyes nor accepted any food or water. I went home at the end of the day and called Tiff. Maybe Dad wanted, or needed, to be alone when he passed. She sobbed. We disagreed about Dad needing to be alone. She said that he needed us to be there when he died. I didn't challenge her.

THE VISITS: DAY 15

Day 15 of Dad's "visits" was generally quiet. He did eat breakfast and lunch today, propped up in his bed, a pureed mash of oatmeal and eggs for breakfast and a pureed sandwich of God knows what and green beans for lunch. While the pureed foods looked gross, we had each tasted them at various times feeding Dad and they actually were not bad at all. Kind of like a meal that had already been chewed for you, but the tastes remained intact.

Dad also stood unassisted today to allow him to be dressed by the staff, which was also a change from the arduous task of rolling him from side to side to put his day clothes on in the bed. He seemed to have more energy now, perhaps a sign that he was getting more oxygen. Although the doctors were way off on their time line, I still trusted them. Dad still had pneumonia and he was still going to die. The question that still nagged was one of logistics; when will it happen and why is it taking so long?

Dad's Alzheimer's had reclaimed his speech and language again. He had lost most of what he had seemingly gained back the previous two weeks. He rarely spoke anymore, and when he did his words were becoming shorter and simpler. Today he didn't speak at all, until 2:45 in the afternoon.

"I might make it."

I was stunned. All that work and "practice" that Dad did, ethereally and practically, was for naught? The visits by the angels, the journeys to the shining city, the clairvoyant statements Dad made? And the doctors' steadfast insistence that he would not recover, that these were his final moments before his tired old body would succumb to pulmonary infection?

I might make it? No, that can't be.

But I said, then encouraged, just the opposite. "You are amazing Dad! Are you going to recover?"

Dad did not respond. He looked like he was somewhere else. I watched the now familiar electrical jolt go up and down through his torso.

"Dad? Dad? Are you going to die or recover?" It came out like I was inpatient. I was. Weeks of not knowing, of preparing for death each and every day, was wearing on me.

Dad replied cryptically. "I know it now." Maybe that was just a random verbalization not even connected to my question.

"Know what, what is going to happen?"

"I am going to die, I know it now, he told me." Dad pointed into thin air.

"Dad, who told you, one of the doctors or the staff, did

you hear them talking about you?" I knew that wasn't the answer, but part of me was still tethered to physical reality.

"No, no, no!" Now it was Dad who was sounding impatient. "Him!" Dad pointed directly at nothing, just the empty space in front of him.

"Your guide? The man with the grey beard?"

"Of course! You know him too!"

"Do you know when you are going to die?"

"Not anymore."

"Does he know?"

"Yes "

"Is it OK?"

"Yeah, its good!"

"Are you sure? So your recovery is temporary?"

Then I left the false security of my earthly tether. "Dad, I know that God has prepared a place for you. I am glad for you."

Dad simply smiled. He then closed his eyes and apparently drifted off, asleep. Although he woke up again later, to eat and to listen to some music I put on, he would not utter another word the rest of the day.

THE VISITS: DAY 16

I entered the room to find Dad, eyes wide open, staring into space across the room. Asimota brought in a breakfast tray. What was once an omelet with toast and potatoes had been reduced to just a large bowl of mush. Dad took in the first spoonful, and it seemed to re-connect him to the physical realm.

"Oh, its eggs." Dad smiled gently.

I wasted no time. I was inpatient now, and almost a little disappointed. How bizarre, three weeks ago praying for a miracle recovery, now disappointed, yes disappointed, that he wouldn't die!

"So, Dad. Are you going to recover? That's what you said yesterday."

"I hope so." That was the last understandable line for a few minutes. Dad mumbled some incomprehensible utterance for the next few minutes, something about being helped by so many others, but I couldn't really make it out. I put on a Benny Goodman CD for him and continued feeding him his "porridge."

"That's my favorite song, there. Tangerine!" Dad was speaking clearly again.

"Mine too." It really wasn't but I was trying to get some sort of reciprocal conversation going. "So is today the day of your recovery?" Impatient again. I was pushing.

Dad seemed to sense my frustration. He gestured with his index finger and shook it.

"Not so fast." Dad was shaking his index finger, like a parent would do with their child, which is exactly what was happening.

"I want to go...I want to go outside today." He tried to position himself to get out of bed, turning to his left and dangling his feet over the edge. For a moment I thought he might stand on his own. He did.

Dad saluted me. "All sirs!"

I then saluted back. "Outside today, yes sir, aye-aye sir! Anchors aweigh?"

"Anchors aweigh my boy, anchors aweigh." A tear traced down my face from the corner of my eye. I helped Dad get his robe and slippers on. I locked arms with him to walk. He pushed my arm back, gesturing with what looked like a baseball "safe" sign from the umpire. I took that to mean I am going to walk on my own.

"Dad, I think I will spend the night over here with you tonight. I still have the feeling something big, very big, is going to happen."

Dad started crying, at first softly with a quivering lower

lip, then more loudly and with some force. But he was smiling while crying.

"Yes! Something big! I know now!" Dad pointed to his watch.

"Is it about time?"

"Yes. I am talking right now with…." Dad's speech became garbled again. It sounded like he not only saw his guide along side us now, but also was communicating with him.

"How much time do you have."

Dad shook his index finger again. "Some time, but not much."

Then Dad added something that might have been important if I had caught it. "Not much time **here**." Dad emphasized the word "here," but I missed the connotation.

Asimota entered the room with a wheel chair. "It's wonderful to see your father standing and getting ready to walk again! Walter is feeling so wonderful today because he has a son that loves him so much! I know you want to walk, sir, but safety first. Let me take you to where you want to go!"

"He wants to go outside, Asi."

"Very well, then," she replied, saluting Dad, who promptly and rather formally saluted back.

"Here, Asi, I can wheel Dad out. We can take it from

here, thank you!" We helped Dad sit down into the chair, Asi put his feet up on the foot rests, and I wheeled him out of his room and into Claire Bridge's circular floor plan, en route to the "veranda." The veranda was an inner garden and patio surrounded by the building, so it was impossible to wander off or get lost.

I had no intention of wheeling Dad into the veranda. He wanted to walk unassisted today, and by God, he was going to. I stopped the wheelchair by the door to the veranda, put the brakes on the chair, and pulled Dad to his feet. Once standing, he pushed my arm gently off of him. I had made the right decision.

We walked a few laps around the small veranda, a lap being maybe 50 feet, and I noticed Dad was not out of breath and not tired. We did a few more "laps" and then sat down on one of the covered tables, shaded by its large umbrella.

We ended up staying out there, well shaded by the umbrella, all afternoon. One of the staff made a large pitcher of lemonade. We'd have visitors from time to time, staff and some residents, who would come out and offer kind words. I asked for Dad's electric trimmer and cut his hair, a "navy cut" of course, and marveled proudly when I had finished what was a pretty good job.

A couple hours later Tiff came by with several plates of nachos from El Azteco, with Dad's being pureed. Feeding Dad and eating off paper plates had the feel of a summer picnic. No one was feeling sad or distressed today. Soon, Dad's primary physician Dr. Jeffries came outside and sat down next to us. It was his day to make his rounds in the geriatric and memory care facilities.

"I've been talking with your dad's staff, and looking at his vitals."

I cut him off, again stymied by my own impatience. "Is he recovering?"

Dr. Jeffries paused a very long time, obviously trying to select just the right words to answer.

"Your father….is truly amazing." And with that simple, nondescript answer, Dr. Jeffries gently stood up, not wanting to impose. "Enjoy your nachos. El Azteco?"

"Yes. Thank you, Dr. Jeffries."

Dad saluted him, and Dr. Jeffries saluted back. As he was walking away, he turned and looked at Tiff, then me. "He is… simply… amazing."

THE VISITS: DAY 17

I had spent the night in Dad's room, convincing myself that "something big" was going to happen during the overnight. Nothing did, at least nothing meant for me.

This morning Dad was quiet, no speech whatsoever, although he was smiling a lot. His Alzheimer's-induced language affliction was definitely coming back, steadily and insidiously, although he still had some moments of lucidity and clearer speech. We just couldn't really predict, nor expect those moments anymore with any regularity.

Tiff came over around noon, to feed Dad a pureed Whopper and onion rings from Burger King, which he wolfed down. She told me she could hang out with him all afternoon, and for me to go home, get myself cleaned up, and get some rest, which I did.

When I returned to Claire Bridge at 4:30 Tiff had a look of total happiness on her face. "Garrett, I think he's out of the woods, I really do, he doesn't seem to be struggling in any way this afternoon. We went for a walk around the building, he's not out of breath anymore and his hands and feet are warm. They took his cath out today and put him back on Depends."

Tiff continued rapid fire, obviously well caffeinated. "In fact I am so confident I am going to join my family up on Lake Michigan for our family vacation. I think Dad is fine, just fine."

Dad looked like he was steadily getting better for days, of course, but my faith in medicine wouldn't let me entertain what a week before was the impossible. Tiff was the first to just boldly say it out loud.

"You sure, Tiff? I mean, yes, I agree, he looks good and if you want to go, go…. I can keep an eye on him down here for the next week."

"Yeah I think I should go, I haven't seen my family hardly at all since this whole thing started."

"Yeah I know, me too."

"Tell you what. Let me go up north with my family this week, and then you and your family can take off next week. Really, Dad is fine, he's beaten this."

I said OK but wasn't really sure if I agreed. Medicine said he was going to die, and his ethereal visitors, if indeed real, were preparing a place for him in heaven. He had two forces at work, and both culminated with Dad's death.

Tiff hugged me as she was leaving. I hugged her back, starting with an obligatory light hug and some pats on the back. Recognizing the shallowness of this act, I then pulled her back in, held tight, and said softly, "We did good."

Tiff had tears streaming down her face when she finally let go. "Well, I will see you next week. Call me every day, OK?"

"I will."

As Tiff left the room, I heard another voice come from behind me. It was the first words Dad had said all day.

"We did good."

"Yes, Dad, we did…"

Dad was able to walk to dinner that evening unassisted. He was quiet again through the evening. At 8:30 I was back in his room, Dad propped up in his bed for the overnight, and already sleeping. Claire Bridge's part time on-call nurse, Michael, came in to see how Dad was doing without his catheter.

Dad woke up, maybe in response to hearing Michael's voice.

Dad started talking, clearly, looking alternately at me and at Michael.

"I still see the angels. I remember heaven. I chose to come back. Me, I did. For a little while."

Day 17 would be the last day of Dad's ethereal visits, at least the last day that we, meaning anyone around him, were included in his visits. He would never speak of the visits, nor have the jolting experience, again.

THREE MONTHS LATER

I was making my usual, nearly daily visit to see Dad on my way back home from work. Dad was where I always found him of late, walking "laps" along the circular floor plan at Claire Bridge. Although he often had staff or residents with him, this time he was walking alone, and I merged into his path to join him. For about twenty minutes we walked, with Dad saying nothing except for an occasional smile and a salute to people passing by

As Dad's "luck" would have it, Dr. Jeffries walked in to do rounds. I watched as he scanned the room, so I could wave and say hi. Dr. Jeffries' eyes found us. Instead of waving, he bee-lined over to us and joined us in our walk.

"How is he doing, Garrett?"

"He's back to normal. Eating, walking, able to be changed easily, happy, but he has very little language again. He doesn't seem to have any physical distress and seems to have gotten into good condition by walking, like he used to do."

"He is amazing," Dr. Jeffries patted me on the back.

Then I took a risk, kind of like Tiffanie's boldness when she declared Dad well.

"Dr. Jeffries, I would like you to order some x-rays for him, to show his lungs have healed up. You know, that big portable x-ray machine I've seen wheeled in here before…"

"Yes I could, his Medicare could cover that, but we already know what it will find."

"We do?"

"Yeah. He had a quarter lung that wasn't collapsed, remember?"

"Yes, of course I do." I could feel where this was going, and it wasn't the right path.

"Well, you can survive on a quarter lung if you don't exert yourself too much."

"But he is exerting himself. The staff say he is walking all day, everyday. Just like he used to."

"I don't know if I can authorize this Garrett, I just don't know."

I was pissed. I also revered Dr. Jeffries, however. He had been through nearly Dad's entire slide into Alzheimer's with our family, and I didn't want to create any rifts now.

I knew Dr. Jeffries probably wouldn't authorize it, but I didn't want to just back down to the managed care influenced decision. "You will find both lungs clear. That would be important to know."

Dr. Jeffries just smiled and broke from the walkers' club. I had pinned him. I was proud of standing up for my dad, and myself, but there was no way Dr. Jeffries would believe my words. At least I had gone down swinging.

The next day I was over at Claire Bridge, again doing laps with my dad. To my surprise, a portable x-ray machine was wheeled in while I was there. I assisted Angela, the main Claire Bridge nurse, and the x-ray technician getting Dad positioned. The chest x-rays were taken. I asked the technician how they looked, but he said only a doctor could read them for us.

The following week Angela gave me a call, asking me if I could come in that day and review the x-rays with Dr. Jeffries. I said sure and met Dr. Jeffries at 4 pm that afternoon. Although I was ten minutes early, Dr. Jeffries was already there when I arrived, talking with Angela.

They motioned me to come into Angela's office, with big smiles on their faces. Dr. Jeffries gave me no time to become impatient.

"The results of the x-ray show both lungs clear, healthy, and fully inflated. You were right!" Dr. Jeffries said this as if he was joining in a celebration, even joyful in the news that he had been wrong.

"I knew it!" I was giddy with joy. "But, how do you explain it?"

Dr. Jeffries paused like he often did, searching for just the right words.

"I would call this miraculous, wouldn't you?"

"Yes I would, Dr. Jeffries, yes I would."

"Out beyond the ideas of wrongdoing and righteousness is a field. Let me meet you there. When the soul lies down in that grass, the world becomes too big to talk about."

—*Jelaluddin Rumi*
13th century

EPILOGUE

My father, Walter Turke, MD, U.S. Navy man to the end, passed away nearly six years later, on June 22, 2013. We were all by his bedside when he could no longer breathe with any regularity. His hospice doctor, Dr. Lopez, one of the most compassionate doctors I certainly have ever known, asked us to trust him and to leave Dad alone so he could make his "transition" without distraction. We all left the room to wait in the Claire Bridge living area

A few moments later the facility's therapy dog, Lilly, trotted into Dad's room, then came back and stood before us. This may or may not have happened, but Tiffanie and I still swear Lilly nodded to us when she came back. We all walked back over to Dad's room. His body was there in his bed, but it didn't even vaguely remind us of Dad. It seemed like just an empty shell. Dad was gone. He had passed over, peacefully, with his entourage of angels.

They say that belief is more powerful, and harder to impact and change, than science. Science has often prided itself that it is not tethered to belief, that it is fact-based, and that it is the correct path to understanding.

I heard once that modern science estimates that the visible universe is only .01 of the total, as we know it. What we can see that is tangible and available to our senses is really composed of atoms that are virtually invisible, in constant motion, and surrounded by an "empty" space of seemingly infinite vastness that we know very little about. It is our senses, and our brains' interpretation of these inputs, that makes us believe that something is tangible, "solid," and "real." Buddhists and scientists agree that it is only our consciousness that defines, and gives parameters to, what we call "reality." The great Turkish-Afghani poet Jelaluddin Rumi, and the Indian prophet Gautama Buddha say that in this emptiness we will find pure consciousness, and our hope for bliss. It is in this vastness from which we come, and it is into the vastness to which we will return.

We already know that there is an all-intelligent energy field that binds everything in the known universe. Physicists might tell us that everything is bound by this field, that it is composed of energy that obeys natural laws, and that this energy cannot be created or destroyed, it can only change forms. Christians, Jews, and Muslims, among other "believers," might tell us that God is eternal, all-intelligent, binds all of us interdependently, is omnipresent and omnipotent, and can manifest in different forms. Would I be too bold to say we are all talking about the same thing?

If we are in sync with this all-powerful energy, we glide effortlessly, and good things will come to us. It is when we consciously disconnect from this energy that we become prey, to violence, chaos, insecurity, and our own arrogance and ignorance.

THERE'S A WINDOW TO HEAVEN

As someone of humble origin, once closed by the limits of science and now reawakened to the possibility of the infinite, I have come to realize that the paths that lead are, with certainty, not always tied to what we think we know....

Garrett L. Turke, Ph.D.
May 2018

There's a Window in Heaven

There's a window in heaven
Where the sun lights through
It lets me see a time
When I was walking with you

You can see our life past
From the Promised Land
I'll meet you at the gates
Then you'll take my hand

Memories of your smile
All I need for this time given
So grateful for our time together
When we both were living

There's a window in heaven
Where the sunlight comes through
It lets me see the time
When I was standing with you

We're only here for a moment
The hourglass always turns over
Time's just an illusion
Until we all cross over

I will meet you in fields
Of lavender purple and blue
I will take your hand
This much I know is true

There's a window in heaven
Where the sun lights through
It lets me see the time
When I'll be walking with you

Words and Music by Jordan Horan and Garrett Turke
Copyright 2017; U.S. Copyright Registration # 1-5337309801